Presenting SCIENCE to the PUBLIC

The Professional Writing Series

This volume is one of a series published by ISI Press. The series is designed to improve the communication skills of professional men and women, as well as students embarking upon professional careers.

Books published in this series:

Communication Skills for the Foreign-Born Professional
   by GREGORY A. BARNES

The Art of Abstracting
   by EDWARD T. CREMMINS

How to Write and Publish a Scientific Paper
   by ROBERT A. DAY

A Treasury for Word Lovers
   by MORTON S. FREEMAN

Presenting Science to the Public
   by BARBARA GASTEL

How to Write and Publish Papers in the Medical Sciences
   by EDWARD J. HUTH

How to Write and Publish Engineering Papers and Reports
   by HERBERT B. MICHAELSON

Presenting SCIENCE to the PUBLIC

Barbara Gastel, M.D.

iSi PRESS®

Philadelphia

Published by

iSi PRESS ®A Subsidiary of the
Institute for Scientifc Information®
3501 Market Street, Philadelphia, PA 19104 U.S.A.

© 1983 ISI Press

Library of Congress Cataloging in Publication Data

Gastel, Barbara
    Presenting science to the public

    (The Professional writing series)
    Bibliography: p.
    Includes index.
    1. Communication in science—Handbooks, manuals, etc.
    2. Science—Information services—Handbooks, manuals, etc. I. Title. II. Series.
Q223.G38 1983   501′.41   83-13042
ISBN 0-89495-028-2
ISBN 0-89495-029-0 (pbk.)

Printed in the United States of America
90 89 88 87 86 85 84 83   8 7 6 5 4 3 2 1

Contents

I. Presenting Science to the Public: General Principles

II. Presenting Science Through the Mass Media

III.   Presenting Science Directly to the Public

"One thing I'll say for us, Meyer—we never stooped to popularizing science."

Preface

Contrary to the view expressed in the cartoon, presenting science to the public is an opportunity to which the scientist should rise. It can help the public, it can enrich our culture, and it can aid science and scientists.

To make sound decisions for themselves and their families, community members need information on science, a field that this book defines broadly to include technology and medicine. Scientific information also is necessary in formulating public policy. And the public, as a major funder of science, is entitled to learn about it.

Indeed, the public is eager to hear about science. Both in newspapers[61] and on television news programs,[91] science stories are among the items that interest community members most. In news magazines, issues with cover stories on science boast top sales.[73] Several science magazines founded since the late 1970's are flourishing, and longer-established counterparts also thrive (Table 1). Science museums and related facilities command 150 million visits per year (more than baseball, basketball, and football games combined), and the annual attendance at the National Air and Space Museum rivals that of Disney World.[86] Many lectures and books on science also draw large audiences.

Such interest is fitting, for science is an integral part of our culture. We would not tolerate locking the musically untrained out of concert halls; we would not banish all but art historians from art museums; nor would we restrict Shakespeare and Dickens to scholars of drama and literature. Rather, we have program notes, gallery tours, popular lectures, and more. The public deserves similar help in understanding and appreciating science.

Presenting science to the public can also aid us as scientists. Within our work, nearly all of us must sometimes present science to administrators, students, patients, and others outside our specialties. Efforts to communicate broadly can cross-fertilize fields of science. Helping the public to understand science is likely to foster public support, as well as

TABLE 1. *Circulation of selected science magazines*

Title of magazine	Approximate circulation*
American Health	450,000
Discover	850,000
The Harvard Medical School Health Letter	300,000
High Technology	300,000
Science Digest	570,000
Science 83	700,000
Science News	165,000
Scientific American	700,000
Technology Illustrated	300,000
Technology Review	70,000

* As of 1983 or late 1982.

attract people to scientific careers. Last but by no means least, presenting science to the public can be enjoyable.

We have many good reasons to popularize science, and science can be presented responsibly and skillfully to the public only if scientists take an effective role. Yet most of us lack background in popular communication and have little time to learn about this field. *Presenting Science to the Public* is therefore a brief, practical handbook for busy scientists. It begins by discussing basic principles, then focuses on presenting science through the mass media, and concludes with chapters on conveying science directly to the public both orally and in print. Individual chapters are easy to consult, and the entire guide is quick to read.

May you use this book often, and with pleasure and success.

Acknowledgments

Many individuals contributed directly or indirectly to this book. Journalists, public information specialists, and others shared their experiences and insights. Robert A. Day of ISI Press provided guidance throughout the project and selected peer reviewers, who in turn offered extremely useful remarks on the preliminary manuscript. Gregory A. Barnes, William Bennett, Bill Bryant, Joyce Gastel, Lynn Hall, Wendy Horwitz, Joel Howell, David R. Lampe, Abram Recht, and Carol L. Rogers also commented thoughtfully and helpfully on the draft. Editors Estella Bradley and Maryanne A. Miller saw this book through the final stages of publication after I left for a year of teaching in China.

To all of the above, as well as to others who aided and encouraged me in this project, I express my gratitude. I also acknowledge those who enabled this effort by helping me to combine science and writing in my education and career. Most of all, I thank my parents, Sophie and Joseph Gastel, for all that they have done for me; to them I dedicate this book.

PART I

PRESENTING SCIENCE TO THE PUBLIC: GENERAL PRINCIPLES

Whenever you present science to the public, adherence to the same basic principles will aid in communicating effectively. Therefore Part I of this book is a set of general guidelines that pertain to any setting and medium. The rest of the volume then discusses various situations in which to apply this advice and offers recommendations specific to each.

Chapter 1

Communicating with the Public: Some Basics

In 1947, Paul W. Merrill of the Mount Wilson Observatory issued three principles of poor writing.[58] The first: Ignore the reader. The second: Be verbose, vague, and pompous. The third: Do not revise. The implied advice on communicating well is especially important in presenting science to the public, and it pertains not only to print but also to other channels.

Analyzing the Audience

As Merrill implied, awareness of the audience is basic to effective communication. "It is an excellent exercise to periodically place oneself in the position of the person he is trying to communicate with," says Dennis S. O'Leary, the physician who served as the George Washington University Hospital's spokesperson after President Reagan was shot. "That person may have different concerns, different needs, and different responsibilities. It helps to understand if one is trying to be understood."[62]

"The public" encompasses many populations, each with a distinct spectrum of backgrounds and interests; it includes both "attentives" (the fifth or so of the general population particularly interested in and informed about science) and "nonattentives."[90] Consider the viewers of the local evening news; the readers of *Good Housekeeping*, *The New Yorker*, and *Discover*; and a chapter of the American Cancer Society. Each of these groups wants some kinds of information on science, technology, or health. But the typical member of each may differ in many ways.

When they are interviewing scientists for publications or broadcasts that serve specific segments, reporters usually gear the questions accordingly. Perhaps, for example, a team has developed a new vaccine. All of the journalists may ask basic questions about the product, its history, and its projected impact. Then, however, a local reporter may focus on hometown researchers' roles, a business correspondent may ask about economic implications, and a writer for a women's magazine may inquire about effects on children's health. If you are unfamiliar with a publication or broadcast for which you are being interviewed, ask about its audience; the answer may suggest types of information to present and ways to present it.

When you prepare to present science directly to the public, again consider the audience. Review copies of magazines for which you wish to write, and perhaps discuss the readership with the editor. If you are invited to address a lay group, ask about the membership.

Once you are familiar with an audience, you can adjust your emphasis and mode of presentation accordingly. Whatever the audience's interests and level of sophistication, though, concern and respect are worth conveying. As budding journalists are often told: never overestimate your audience's knowledge, but never underestimate its intelligence either.

Building on the Audience's Background

After analyzing the audience, use its background and interests as a scaffolding for your information. Start by relating your topic to matters about which the readers or listeners already know and care (for instance, the cost of energy, the weather, their hobbies, or their health). Otherwise, the audience may not understand what you are saying—and even if it does, it may see little reason to find out more. Once you have shown how your material relates to the familiar and the practical, you can move ahead; those being addressed will then be able to, and will want to, follow along. Be careful, however, not to progress too far from the original concerns or to move too fast.

So, how to begin? With rainbows, not refraction. Influenza, not immunology. Your public's pocketbooks, not your doctoral dissertation. Consider opening with a historical observation or an item from the news. Perhaps better yet, start with some human interest, such as an anecdote about a famous person, the story of a client or student or patient, or a glimpse of the challenges faced in your own research.

An especially helpful tactic for relating unfamiliar to familiar concepts is the analogy. Comparing a computer to a brain (or vice versa, depending on the audience), a chemical procedure to a culinary technique, or a space probe's sensors and appendages to the parts of an organism can help your audience to grasp and appreciate your point. One example of how an analogy can help make an abstract principle seem concrete is this passage from *The Soul of a New Machine*:

> Computer engineers call a single high or low voltage a bit, and it symbolizes one piece of information. One bit can't symbolize much; it has only two possible states, so it can, for instance, be used to stand for only two integers. Put many bits in a row, however, and the number of things that can be represented increases exponentially. By way of analogy, think of telephone numbers. Using only four digits, the phone company could make up enough unique numbers to give one to everybody in a small town. But what if the company wants to give everyone in a large region a unique phone number? By using seven instead of four digits, Ma Bell can generate a vast array of unique numbers, enough so that everyone in the New York metropolitan area or in the state of Montana can have one of his own.[47]

Constructing such an effective analogy can be difficult, but the effort can be most worthwhile.

In summary, begin by relating your subject to that which is familiar and important to the audience. Once that link is established, you can move ahead. Take care, though, not to stray too far for too long.

Stating the General Concept Before the Details

To orient the audience, and to help it decide whether to find out more, state general concepts before giving details. Give the function, the principle, and the main advantages of an invention before getting to the particulars. State the outcome and main phases of a process before launching into a step-by-step account. Present your major findings and their implications before discussing your evidence at length.

Consider this paragraph from the popular essay "The Panda's Thumb" by scientist Stephen Jay Gould:

> The panda's "thumb" is not, anatomically, a finger at all. It is constructed from a bone called the radial sesamoid, normally a small component of the wrist. In pandas, the radial sesamoid is greatly enlarged and elongated until it almost equals the metapodial bones of the true digits in length. The radial sesamoid underlies a pad on the panda's forepaw; the five digits form the framework of another pad, the palmar. A shallow furrow separates the two pads and serves as a channelway for bamboo stalks.[42]

Had Gould begun by discussing radial sesamoids and metapodial bones, he might have lost most of his audience. By presenting the general concept first, he orients readers so well that they can readily understand these terms.

Including—and Following—Examples

After presenting a general concept, you can support and clarify your ideas in various ways. One effective technique is to use examples, as physician Lewis Thomas does in this passage from *The Lives of a Cell*:

> Most bacteria are totally preoccupied with browsing, altering the configurations of organic molecules so that they become usable for the energy needs of other forms of life.... Some have become symbionts in more specialized, local relations, living as working parts in the tissues of higher organisms. The root nodules of legumes would have neither form nor function without the masses of rhizobial bacteria swarming into the root hairs, incorporating themselves with such intimacy that only an electron microscope can detect which membranes are bacterial and which are plant. Insects have colonies of bacteria, the mycetocytes, living in them like little glands, doing heaven knows what but being essential. The microfloras of animal intestinal tracts are part of the nutritional system. And then, of course, there are the mitochondria and chloroplasts, permanent residents in everything.[85]

Speaking of examples, imitation can be not only flattery but also a good strategy, both in developing general communication skills and in performing specific tasks. Read, listen to, and watch popular presentations of science. If you like an item or it has been acclaimed, consider what makes it good. If you dislike it, identify what you dislike. And when, for instance, you will appear on a talk show, write a popular article, or give a speech, try to find a model to use.

Among the most accessible examples of good science communication are the magazine articles that have won AAAS-Westinghouse Science Journalism Awards. Recent winners are listed in Table 2.

Depicting Relationships Clearly

To you as a scientist, it may be obvious that A implies B, that step C must precede step D, and that E seems to contradict F but really does not because of G. To a reporter or a member of the public, however, such relationships may be far from clear.

Scientists familiar with a topic can easily skip between premises and

TABLE 2. *Models of outstanding science writing: Magazine articles winning AAAS-Westinghouse Science Journalism Awards (1975–1982)*

Brodeur P.   Annals of chemistry: inert. The New Yorker 1975 April 7.

Eberhart J.   Series of 25 articles on U.S. Viking mission to Mars. Science News 1976 June–November.

Cooper HSF Jr.   A reporter at large: life in a space station. The New Yorker 1976 August 30 and September 6.

Bennett W, Gurin J.   Science that frightens scientists: the great debate over DNA. The Atlantic Monthly 1977 February.

Golden F.   Those baffling black holes. Time 1978 September 4.

Canby TY.   Early man in America. National Geographic Magazine 1979 September.

Gilmore CP.   After 63 years, why are they still testing Einstein? Popular Science 1979 December.

Cooper HSF Jr.   A reporter at large: shuttle. The New Yorker 1981 February 9 and 16.

Ferris T.   Beyond Newton and Einstein: on the new frontier of physics. The New York Times Magazine 1982 September 24.

conclusions, but others may need to be told the intermediate steps. For instance, when addressing clinicians, saying little more than "carotene-mia-lycopenemia" may suffice; the audience quite likely can deduce the cause and the result. For the public, however, one must specify the background and the reasoning, as *New Yorker* writer Berton Roueché does in the medical detective story "The Orange Man":

> Dr. Wooten closed the volume. Turner was not only a heavy eater of carrots. He was also a heavy drinker of tomato juice. Carrots are rich in carotene and tomatoes are rich in lycopene. Carotene is a yellow pigment and lycopene is red. And yellow and red make orange.[74]

Various words and phrases also can help the audience to perceive links among ideas. Consider the following paragraph by Isaac Asimov:

> If tubes are fine enough, the adhesive forces are enough to pull the water level a considerable way upward against gravity. Stick a blotter into water, **for instance,** and the water will soak upward through the fine spaces between the matted cellulose fibers of the blotter. Water will **also** soak upward through the very fine tubes leading up tree trunks, and this is one of the mechanisms by which trees lift water hundreds of feet without a pumping mechanism like the animal heart. **Because** this works best in tubes that are as fine as hairs, or finer, **this** lifting of fluids is termed capillary action or *capillarity* from the Latin "capillus " (hair).[4] (boldface added)

Without the words denoted by boldface, this passage would be much harder to follow. And without other such transitional terms and phrases (such as "then," "therefore," "in addition," "however," and "nevertheless"), so would many popular accounts of science.

Tactics other than words also help the audience keep its place. In written items, such structural devices as headings, italics, and numbered lists often enhance clarity. Similarly, in speech, skillfully timed pauses, as well as changes in volume or tone, can aid in orienting listeners.

Avoiding Jargon

Jargon-slinging may well be the most widely caricatured behavior of scientists. But when you are trying to present science and the public is trying to understand, the effect is hardly funny at all.

DOONESBURY   **by Garry Trudeau**

Be alert for jargon, which may be so entrenched in your vocabulary that you normally do not notice it. Avoid technical terms unless reason exists to acquaint the audience with them. For instance, say "aspirin," not "acetylsalicylic acid"; "thermos bottle," not "Dewar flask"; "plants," not "vegetation" or "flora." Let the audience concentrate on your ideas rather than puzzle over your words.

In particular, beware of presenting ideas in mathematical terms. Most members of the public have little training in math, and many of the rest have forgotten what they learned and are intimidated by the field. Although science journalists often can help to translate other material

© 1980 Sidney Harris
Created for *Discover*

into popular language, they are usually of little aid here. As one veteran science writer puts it, "When scientists march over to the blackboard and start putting formulas on it, I'm through."

When a term is defined differently in your field than in everyday use, make the distinction clear. For instance, explain that "depression" can denote a serious biological illness in addition to meaning "the blues," and distinguish what "thinking" or "memory" signifies in a computer from what it means in human beings.

THE FAR SIDE   By GARY LARSON

"Hey! What's this Drosophila melanogaster doing in my soup?"

Introducing New Terms Gently

Technical terms need not be avoided totally. Some will be of lasting use to the audience. Some are important in explaining your material. Perhaps an occasional such word helps lend authority to what you say. Beware, however, of introducing technical terms too abruptly. Doing so may perplex, and so drive away, the public that you are trying to reach.

For an excellent model of leading up to a scientific term, note how Sharon Begley begins a *Newsweek* article:

> Magnetic poles, like bookends and honeymooners, always seem to come in pairs. For every north pole there is a south, and trying to divide one from the other is an exercise in futility. Cutting apart magnets doesn't produce separate chunks of north and south, but miniature replicas of the original. Nevertheless, in 1931 British physicist P.A.M. Dirac theorized that, just as there are single particles of electricity, there must also be "magnetic monopoles."[8]

Likewise, observe how writer Bernard Asbell skillfully introduces a technical phrase:

> ... Snell discovered a single cluster of genes exercising central command over the "decisions" of the others. Some scientists have fondly dubbed this cluster "supergene." More formally Snell calls it the "major histocompatibility complex."[3]

Presenting an idea in simple language before assigning it a name can allow you to introduce technical terms without alarming, and thus repelling, the audience.

Keeping Words, Sentences, and Paragraphs Short

Your audience faces the challenge of understanding and assimilating information that is unfamiliar and technical. It should not also have to contend with lengthy words (such as those in the preceding sentence and many others that scientists frequently use), convoluted sentences (such as this one, with its parenthetical comments), and rambling paragraphs.

Whether you are addressing the public directly or being interviewed by a journalist, keep most of your words, sentences, and paragraphs short; Table 3 lists brief substitutes for terms that scientists often use. Such brevity is essential for presenting science to general audiences.

TABLE 3. *Examples of using shorter, more common words*

Long word	Short word(s)	Long word	Short word(s)
approximately	about	necessary	needed
attempt	try	numerous	many
complicated	complex	opportunity	chance
concerning	about	optimal	best
consequently	thus	perform	do
construct	build	possess	have; own
currently	now	previously	before
demonstrate	show	primarily	mainly
elucidate	explain	principally	mainly
essential	needed	provide	give
fabricate	make	subsequently	later
fundamental	basic	sufficient	enough
illustrate	show	superior	better
indicate	show	technique	method; way
initial	first	terminate	end
initiate	begin	uncommon	rare
instrument	tool	unnecessary	needless
manually	by hand	utilize	use
modify	change	verify	prove
multiple	many		

Granted, longer elements, such as those in the above passage by Lewis Thomas, can be appropriate for the more sophisticated segments of the public. Even the more educated, though, usually appreciate having material presented in an easily digestible way.[37,45]

If you are preparing a document for the public, you can estimate how easy it is to read by using a formula that takes both word length and sentence length into account. One such formula, which yields a quantity called the Fog Index, is discussed on pages 103 through 106.

Using Numbers (But Not Too Many)

The public expects, and tends to remember,[25] statistics in discussions of science. "How big? How fast? How far? How many? What percentage? And how expensive?" readers or listeners may well want to know.

Provide such information, but beware of overwhelming the audience with numbers. Also, put quantities in a form and context meaningful to those you are trying to reach. For example, avoid scientific notation, statistical terms, and unfamiliar units. And do relate magnitudes to those familiar to the public. A quick, unobtrusive comparison (such as between an extinct creature's size and a zoo animal's, or between the probability of event X and the likelihood of winning lottery Y) works best. When, however, you want to emphasize a quantity, consider constructing a more extensive analogy, as is done in the excerpts below (drawn, respectively, from a report in *The New York Times* and an editorial in *The Boston Globe*):

> Thirty femtoseconds is the shortest event ever perpetrated by man. Technically, it is .03 picosecond, which is one trillionth of a second—but Charles V. Shank, head of the Bell team that built the pulsed-laser device, has a better way to describe it. "In one second, a light beam can travel from the Earth to the Moon and back," he explained. "In 30 femtoseconds, light travels about 10 microns, or about one-tenth the thickness of a human hair."[33]

> Nothing is perhaps so impressive as the elegant accuracy of the [Voyager 2] shot itself. In four years of travel since its launching, Voyager had traveled more than 1.2 billion miles, passing Jupiter and its moon system before arriving at Saturn within three seconds of the scheduled time and passing about 30 miles from its theoretical aim point.
>
> It is hard to find an earth-bound analogy, but the shot is rather like a golfer teeing up in Boston, hitting a drive that soars across the country. It hits exactly the right freight car in a speeding train in North Dakota (on the left side, please, not the right) and bounces southwest across the rest of the country and out over the Pacific. Near Hawaii, the ball passes as planned through a predetermined porthole in a tanker steaming south, and then bounces off the deck and off to the beyond.[2]

Providing Illustrations (and Keeping Them Simple)

Some years ago, a scientific illustrator noted with surprise that he had produced some of his most effective work when he was rushed. Upon reflection, he realized why: when time was short, he omitted the extraneous details.[21]

Do use illustrations. A picture is said to be worth a thousand words, and in presenting science to the public its value may be incalculable. Also, in this age of television, the public has come to expect visual aids. But as the story above suggests, keep illustrations simple unless they are meant only to attract attention or to decorate.

One role of illustrations, of course, is to portray physical objects. If realism is important, the best choice may well be photographs. Otherwise, diagrams, which allow you to emphasize pertinent features and eliminate the rest, are often the best medium.

For instance, imagine the task of preparing a lay primer on coronary artery bypass surgery. In the booklet, you wish to show that the heart has only a few main arteries. Do you choose a set of x-rays including views such as Figure 1? No, the public could hardly be expected to interpret such illustrations. Figure 2 is more appropriate, but it contains too many extra details. The best choice is Figure 3, which includes little other than the structures being discussed. Similarly, to indicate that the blocked blood vessels can be bypassed surgically, use a schematic diagram (Figure 4) rather than a photograph (Figure 5).

Another role of illustrations is to help convey abstract ideas. Graphs

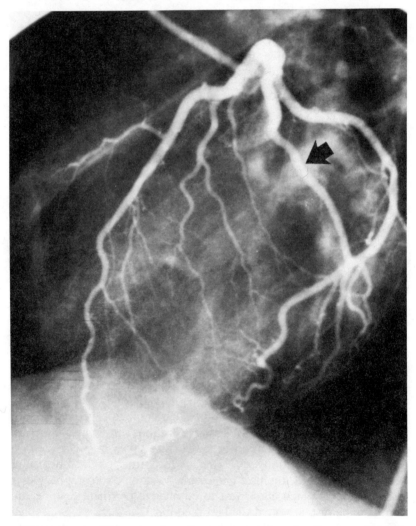

FIGURE 1. *An illustration in a form inappropriate for the general public.* *(Courtesy of David C. Levin.)*

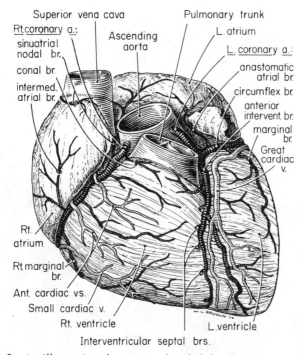

Superior vena cava

Rt coronary a.:
sinuatrial
nodal br.

Ascending
aorta

conal br.

intermed.
atrial br.

Rt.
atrium

Rt marginal
br.

Ant. cardiac vs.

Small cardiac v.

Rt. ventricle

Pulmonary trunk

L. atrium

L. coronary a.:

anastomatic
atrial br.

circumflex br.

anterior
intervent. br.

marginal
br.

Great
cardiac
v.

L. ventricle

Interventricular septal brs.

FIGURE 2. *An illustration that is too detailed for the general public. From* Essentials of Human Anatomy, *Seventh Edition, by Russell T. Woodburne. Copyright © 1983 by Oxford University Press, Inc. Reprinted by permission.*

can aid in presenting quantitative material. Here, too, stay with the simple and the familiar; for example, use bar graphs and pie graphs, but avoid semilogarithmic plots. Flow charts can assist in describing multi-step processes. Avoid those that are convoluted and detailed (e.g., Figure 6); instead, use those that are straightforward and present only the essential elements (e.g., Figure 7).

Photographs, diagrams, graphs, and other aids can add much to your own writings and presentations for the public. And if you are being interviewed for a publication, television program, or film, do not hesitate to suggest visuals that might help to convey your ideas. The journalist may well be groping for ways to illustrate the story and most likely will appreciate your aid.

Repeating Major Ideas

Many of us recall our high school English teachers' telling us: "Say what you're going to say, say it, and then say what you've said." Such

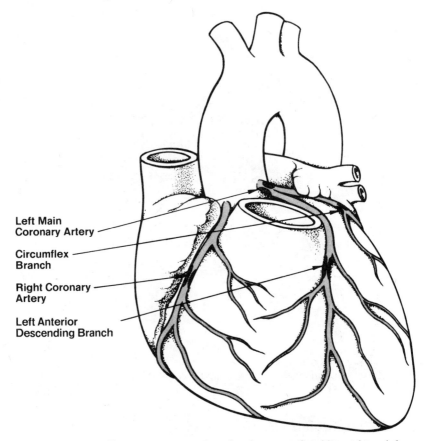

Left Main
Coronary Artery

Circumflex
Branch

Right Coronary
Artery

Left Anterior
Descending Branch

FIGURE 3. *An illustration appropriate for the general public. Adapted from* Modern Medicine *1980 November 15–30:37.*

advice pertains well to presenting science to the public. Unfamiliar, complex material is hard to understand the first time, especially if presented orally. So, state your main ideas, elaborate on them, and then sum up. When along the way you present matter that is difficult but important to grasp, consider restating it in different words.

Checking with the Audience

Well, you have done your best to follow the advice in this chapter. Did the guidelines work? Are you getting your ideas across? The way to find out is to ask. In other words, conclude as you began: by considering the audience.

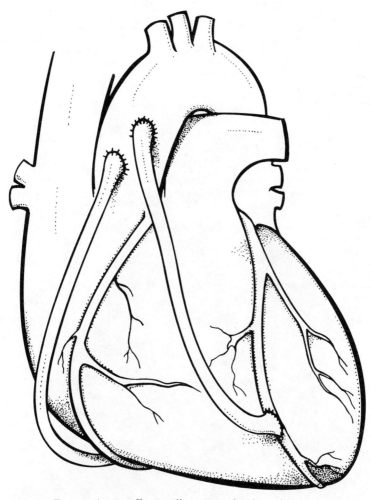

FIGURE 4. *An effective illustration for the public.*

One helpful measure, though one to use with discretion and tact, is to have journalists restate your main points; you can then further explain any material not effectively conveyed. Also, encourage reporters to call you if anything later seems incomplete or unclear, and let them know that you would be glad to check passages for technical accuracy.

When you are preparing an item for the public, try to test it out on representatives of the audience—for example, on relatives, friends, or neighbors. Run through your speech for them, or show them a draft of your article. Does the content seem to keep their interest? Do any points puzzle them? Do their comments seem to show that they grasped the

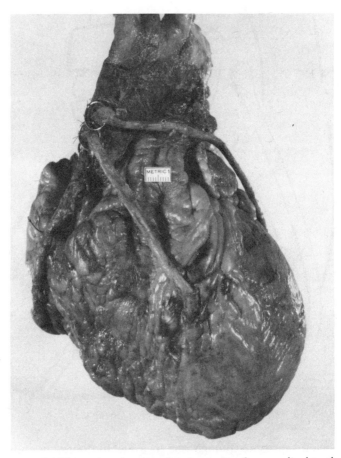

FIGURE 5. *An illustration that may be appropriate for a medical textbook but is not suitable for a publication aimed at the general public. (Courtesy of Maggie Moore, National Heart, Lung, and Blood Institute, National Institutes of Health.)*

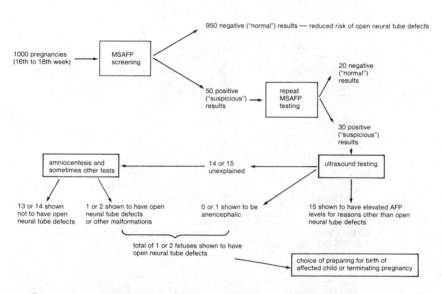

MATERNAL SERUM ALPHA-FETOPROTEIN SCREENING AND FOLLOW-UP TESTS

The above flow chart shows the sequence of tests available in a maternal serum alpha-fetoprotein screening program. Although not indicated above, counseling also is given at each stage. The numbers on the chart are those expected if at each stage all women remaining in the high-risk category choose to undergo further testing.

FIGURE 6. *A flow chart that is needlessly convoluted and is too detailed for the general public.*

FIGURE 7. *An effective flow chart. Courtesy* SCIENCE 83 *magazine.*

central ideas? Do they have questions that were not adequately answered? And do they have suggestions with regard to content or form? Now is the time to find out and, as Merrill's third principle implies, to revise your work accordingly.

PART II

PRESENTING SCIENCE THROUGH THE MASS MEDIA

After their formal education ends, nonscientists gain most of their information on science through the mass media.[56] The media are also an influential source of scientific information for many who establish public policy. Thus, learning to work with the press is important in seeing that science reaches the public and is used on its behalf. As those of us in science know especially well, learning how something works can greatly aid in working with it. Therefore, this section of the book discusses the who's and how's and why's of science reporting, as well as offering advice for presenting science through the mass media.

Chapter 2

Recognizing Who Reports on Science

Who reports on science? A varied lot: general reporters, professional science writers, freelancers, and more. Does the answer matter? Yes. Journalists with various backgrounds can, and do, write well about science. But the kind of information they need, as well as the form of information that they find most useful can differ considerably. Knowing who is covering science, both in general and in any particular case, thus aids in presenting science through the mass media.

General Reporters and Non-Science Specialists

General reporters and journalists with specialized beats other than science present much of the science that appears in the press. The reason is at least twofold. First, science is often a component of general news. Says Fred Jerome, who directs the Media Resource Service at the Scientists' Institute for Public Information: "Most of the science and technology news in this country is related to very important national stories." For example, scientific information often appears in reports on environmental issues, earthquakes, storms, public figures' illnesses, military technology, and the economy.

Second, few newspapers, radio stations, and television stations boast science writers of their own. Of the estimated 100 full-time science writers working on newspapers today, groups are clustered at major dailies such as *The New York Times* and *The Los Angeles Times*, and the rest are scattered elsewhere. Thus, of the approximately 1,750 daily newspapers in the United States, only about 50 have full-time science writers.[64] In broadcasting, science reporters are even less common. To convey science news, the media therefore depend mostly on stories by their non-science reporters and on items from outside sources such as the wire services.

The general reporter writing about science faces various obstacles. Unlike the full-time science writer, such a journalist has many topics competing for attention; when the mayor calls a press conference or a fire breaks out, the science story is likely to get short shrift. If the reporter is in a small town, the nearest authority may be far away, and even an authoritative written source may be hard to find. Then there is the matter of science itself, in which most general reporters are untrained.

General reporters vary greatly in willingness and ability to tackle science. At one extreme are those like the journalist who telephoned Robert M. Byers, who heads the news office at the Massachusetts Institute of Technology, one day. Suddenly having been drafted to cover a local hearing on recombinant DNA, the bewildered reporter asked for a five-minute rundown on the subject. The last words in his request: "But don't tell me anything hard."

At the other extreme are journalists like David W. Crisp of the *Palestine* (Texas) *Herald Press*, whose series on nuclear waste won an AAAS-Westinghouse Science Journalism Award in 1980. Says Crisp:

> Unlike many subjects reporters cover, science writing actually comes easy once the basic principles are grasped and the vocabulary is mastered, because science is comparatively logical, orderly and precise. The study of long term radiation effects holds no terror for someone who has tackled the Texas Property Tax Code, and I would rather debate relativity with Einstein himself than to try to explain Texas politics.[19]

In short, the general reporter, especially if at a small newspaper or station, is likely to lack science training, to have few sources of science information other than interviews, and to be hurried and harried. Thus, the job of presenting science to the public depends largely on the scientist. As Fred Jerome advises, "Explain, explain, explain—but without resentment." The better suited your words are for the public, the greater the chance that the story will be written and published, and that it will convey science accurately and effectively.

Talking with general reporters can take considerable thought and work, especially the first few times. The effort, however, can be most worthwhile. Not only do items by general reporters represent much of the science reporting done. Also, material in small-circulation newspapers and magazines can ultimately command large audiences by feeding into wire services and providing story ideas to other journalists.

Political, business, lifestyle, and other types of reporters also must sometimes obtain and present scientific material. By working effectively with such journalists, you can help to convey science to a wider audience than that which reads science articles. You can likewise aid in showing the public that science and technology are basic to many facets of our lives.

Science Writers

Science writers are a recent and distinctive breed of journalist. Many of the first were "radio editors" who in the 1920s instructed the public in building their own crystal sets and then, when radios came to be mass-produced, turned to reporting on science.[48] The National Association of Science Writers was formed in 1934;[15] its first eleven members were virtually the only full-time science writers in the United States.[63] Then, in 1937, five pioneering science writers (John J. O'Neill, William L. Laurence, Howard Blakeslee, Gobind Behari Lal, and David Dietz) won a Pulitzer Prize for their coverage of the Harvard University tercentenary.

Most of the early science writers came from backgrounds in journalism rather than science. Science writers surveyed several years ago, however, reported a wide range of paths to their careers.[76] Today, many of the younger science journalists have undergraduate or even graduate degrees in science.

How do you identify the science writer? By the size of the mailbox. Unlike most other journalists, science writers receive huge amounts of mail: piles of press releases, stacks of journals, and more. Many science writers also keep exceptionally extensive files.

Another difference is that whereas other journalists often shift specialties or careers, science writers tend to stay with what they do.[27] Perhaps the outward stability stems in part from the work's inherent variety. David Perlman of the *San Francisco Chronicle* has praised his job as "a full-time, perpetual fellowship to a graduate school with an endlessly varied and endlessly challenging curriculum."[63] Whatever the reason for it, this continuity can help scientists to present science through the mass media. A journalist who has long covered your field probably will grasp your ideas readily. In addition, science writers need to maintain good relationships with sources and thus are especially committed to responsible reporting on science.

Several other features of science writers also may be of note to scientists dealing with the press. Especially once well established, science writers often have great autonomy[24]; whether labeled "science editor" or not, they frequently choose their topics rather than having them assigned, and their work tends to undergo little editing. Their deadlines often are looser than those of general reporters.[53] And science writers tend to identify more with each other[24] and perhaps even with scientists[53] than with the journalistic community.

In what they read, where they go, and whom they know, science writers sometimes resemble scientists more than they do general reporters. Especially if they subspecialize, science writers often are well versed in technical language; for example, Kelly Beatty, senior editor of *Sky & Telescope*, says of the magazine's staff: "Among ourselves we talk science, but when we write we write English." Many of them spend considerable time scanning scientific journals. They attend scientific conferences and seminars. They often have extensive files of technical information. They also have networks of contacts in the scientific community.

In fact, one science writer boasts that "I'm often more current on affairs in their discipline than the scientists are." However, science writers are generalists, not practicing scientists. You can usually expect a science writer to understand basic scientific principles and terms, realize how science works, and know how to present science to the public. But you cannot count on those who cover everything from asteroids to hemorrhoids (or even just all of astronomy or medicine) to be intimately acquainted with your own subspecialty. Even if they are, they may begin with very basic questions, both to check information already obtained and to get statements in your own (quotable, they hope) words.

As noted above, science writers tend to identify with each other. Science-communication scholar Sharon Dunwoody has studied this phenomenon and characterized what she has called "the science writing inner club."[24] This group (an influential one, as most of its members work for wire services, large newspapers, or major magazines) coalesced in the 1960s from journalists covering the space program together. Today, when the members congregate at scientific conferences and other events, they still cooperate. In such situations, they often choose the same topics for their stories and cover them similarly.

In many ways, science writers are thus a distinctive group. They know and care about science and scientists. Knowing and caring about them can aid scientists in presenting science through the mass media.

Freelancers

Anyone can assume the title "freelancer," and people with virtually every background do. The voice that identifies itself as a freelancer's could belong to a general journalist, a science writer, or a scientist preparing a popular article. It might also be that of an unemployed (and perhaps unemployable) would-be writer who is competent in neither journalism nor science. And the freelancer may have a firm assignment from an editor or simply hope to sell a piece. How best to gear what you say, and in fact whether to accept an interview at all, depends on who is contacting you and why. Thus, in dealing with a freelancer, it is especially important to find out the caller's background and plans.

As this chapter has described, those who report at least occasionally on science vary greatly in their backgrounds, resources, and constraints. Knowing who is covering science, both in general and in any particular case, can help you present science through the media. To work most effectively with the press, however, you should also understand the way it works. Thus, the next chapter discusses how journalists cover science.

Chapter 3

Realizing How the Press Works

To work effectively in a research institution, hospital, or company, a scientist must understand how such an organization functions. Similarly, knowing how the media operate is basic to working effectively with the press. Thus, this chapter discusses how journalists choose topics and prepare stories; it emphasizes those aspects especially relevant to scientists.

However journalists cover science, they have long been doing so. In 1690, the first (and only) issue of the first American newspaper, *Publick Occurrences*, contained a medical piece beginning "Epidemical *Fevers* and *Agues* grow very common. . . ."[49] Also in the colonial period, Benjamin Franklin often published science essays in the *Pennsylvania Gazette*.[48]

Today, reports on science, as well as other stories with science components, appear in a wide range of sources: newspapers, magazines, television, radio, and more. As Chapter 8 will discuss, each of these media functions somewhat differently and thus demands special consideration. But first, let us look at the basic process of reporting on science. Although some of the material below is phrased in terms only of print journalism, most of the chapter pertains to broadcasting as well.

The Story Idea

Many factors influence what ideas for stories arise, as well as which ones survive. Knowing these factors can help scientists both to seize existing chances for presenting science to the public and to create new ones. In addition, such awareness can aid in anticipating when journalists will call.

Events in the general news sometimes spur reporting on the related science. Thus, if you are a geologist and the ground has been shaking, a meteorologist and it is snowing in July, or a psychiatrist and a major trial involves an insanity defense, a journalist may well call. Because reporters other than science writers may not distinguish among subspecialities of science, they may approach you about areas peripheral to your own.

Stories devoted specifically to science arise from various sources. Widely circulated and wide-ranging journals such as *Science* and *The New England Journal of Medicine* are standard founts of science news. So are major conferences on broad areas of science (for example, the annual meetings of the American Association for the Advancement of Science).

Particularly if they specialize in science, reporters may also obtain ideas for science stories through less widely accessible routes. Science writers scan the more specialized journals. They attend smaller and more narrowly focused conferences, including both technical meetings and "science writers' seminars" (conferences intended to brief science journalists on, and obtain coverage of, areas of science). They visit laboratories. They talk informally with scientists after interviews, at parties, and in conference corridors. In addition, both scientists and members of the public sometimes suggest topics to them.

For both science writers and other journalists, initiatives by scientists and their institutions are key sources of story ideas. Much science coverage originates from press releases, press conferences, and other intentional contact with the media; sometimes reporters use only those sources in covering a topic or event. "The pattern throughout the history of the recombinant DNA issue," writes Rae Goodell of the Massachusetts Institute of Technology, "is one in which the scientists willing to speak out—whether they were the Asilomar conference organizers, the established scientist-critics, the Congressional lobbyists, or the industrial spokesmen—have largely determined the timing, extent, and direction of press coverage."[39]

In science, sources can indeed set the agenda for the media. Elaine Freeman, who heads the Office of Public Affairs at The Johns Hopkins Medical Institutions, describes a case in point: a natural experiment that occurred when a paper on drug testing appeared in the *Proceedings of the National Academy of Sciences*. Because of the scientists' travel schedules, a press briefing took place not when the paper appeared, as normally happens, but two weeks afterward. In the interim, Freeman recalls, her office received only one call about the paper, from a reporter for *Science News*. However, she notes, "When we finally held the briefing, reporters attended from *Time, The Washington Post*, and the television networks, among others. Even the FDA learned about the new test not from PNAS but through calls from reporters who attended our press conference."[36]

Furthermore, journalists get ideas from each other's stories. For example, findings may first reach the public through an item in a newspaper. A magazine reporter may read that article and then file one on the topic. Other journalists may then decide to cover the work too. In short, a chain reaction can occur. In fact, once coverage exceeds a certain threshold, more is likely indeed. If a science story makes it to the front pages of *The New York Times* and *The Washington Post*, editors of news magazines may feel obligated to run an article on the topic, regardless of whether their science writers consider it important and interesting.

In this case and others, writers and editors can disagree on a story idea's worth. Some editors have all the technical sophistication of the one who, when the first news of Three Mile Island came over the wires, called the science reporter and said, "They mentioned something called a meltdown. Is that something we should be concerned about?" Thus, what the science writer finds exciting sometimes seems too hard or esoteric to the editor. And when the editor suggests a topic, the science writer may find it of little scientific novelty or significance. "I'm the one who knows what's going on in science," the writer may protest. "But I have a better sense of what interests the readers. You're too much a specialist to be in touch with that," the editor may retort. Each has a valid point.

Newsworthiness

The above debate revolves around an elusive question: what gives a subject press appeal? Recognizing some of the answers can aid in anticipating calls from journalists. It also can help in clothing your message in newsworthy terms when you are hoping to disseminate information through the mass media.

Newsworthiness frequently depends on the proper balance of familiarity and novelty. The conveyers and consumers of news often pay the most attention to items pertaining to their existing concerns. Yet unless an item differs enough from what they already know, it will not hold their interest; without being new, something is not news.

Various aspects of a topic can give it a press-drawing pertinence. As a booklet issued by the American Chemical Society states, "Items that affect the welfare, health, or wealth of a large number of people . . . will find ready acceptance as news."[16] Relevance to basic drives, too, can make the subject a sexy one. Geographic proximity is also a factor: your neighbor may be much more interested in your research than in studies by a far-away scientist. And members of the public care about what they encounter in their daily lives; when he was science editor of *The New York Times*, William Stockton noted that "A story about cockroaches that appeared on the front page of 'Science Times' turned out to be one of our most popular stories and evoked more mail than anything we've run."[87]

Human interest can provide the familiarity that attracts journalists and their audience. Elaine Freeman of Johns Hopkins has observed that sometimes reporters, especially those for television, will not attend a press conference unless a patient will be there.[36] Because people also care about neighboring species, almost-human interest can likewise add to newsworthiness.

Yet to command attention in the media, an item must be unfamiliar enough. For instance, something must change. As Sharon Begley, science editor of *Newsweek*, puts it, "News comes in the unit called events."[7] Having just been discovered, or announced or published today, can help make an item news. Freshness, however, can be in the eye of the beholder. Even a finding made or a technique developed months or years ago can seem novel to those unaware of it.

Surprise and wonder can also make a subject special. "I like to do stories that will lift people out of themselves for a few minutes," says Begley. "We don't always do stories about coffee and cancer, or about nuclear waste. We do stories about astrophysics and black holes and embryonic development, which seems to be nothing short of a miracle."[7]

Boyce Rensberger, a senior editor of *Science 83*, expresses similar sentiments. "The 'gee whiz' story, as detractors like to call it," he states, "is . . . the heart of science writing. It's the reaction every scientist would like to elicit from colleagues through a professional paper. 'Gee whiz,' after all, is an expression of fascination and appreciation felt upon learning of some new and compelling piece of knowledge. That's what science is really all about."[71]

Other factors can also lift subjects from the ordinary into the newsworthy. Conflict, controversy, and contest, as well as danger, are traditional elements of news. Suspense and adventure, including those entailed in research, attract the press. Prominent people cast an aura of newsworthiness on much around them; the most mundane of disorders draws attention when it afflicts the President or sidelines a famous athlete, and much that Nobel Prize winners say and do is considered news.

Whether an item makes it into the news depends in part on how easily and briefly the journalist can present it. "Unfortunately," Rensberger has noted, "if it takes a lot of words to prepare the reader to be fascinated, the story won't be considered newsworthy."[72]

In addition, timing influences whether a story is conceived and born. Although several papers at a conference may be newsworthy, a reporter may have time to cover only one or two; a journal article that might receive media attention in a slow week may fail to do so if it has to compete with a space mission. Also, an event occurring too late to make it into the next edition but too early to seem fresh for the following one may never appear in the news. Thus, awareness of competing events and journalists' deadlines is important in planning activities such as press conferences.

Even if the reporter has time to prepare a story, space (or, in broadcasting, time) may not be available to run it. Science writer Robert Cooke of *The Boston Globe* laments, "Every time Teddy Kennedy gets the hiccups, science gets kicked out of the paper." And one summer at *Newsweek*, the editors yanked a story on cancer and diet shortly before the magazine went to press. The reason: Not enough space. There had been an upset at Wimbledon. The next week, however, the article did appear.[17] Fortunately, the story was a feature rather than a news article and thus was not highly perishable.

Types of Stories

As implied above, science can appear in various classes of stories. The type of article produced depends in part on what made the topic newsworthy. If the newsworthiness stems largely from newness, a *news story* is likely to result. Such an item may begin: "A pulsar throbbing 163 times a second has been discovered between the constellations Vulpecula, the Little Fox, and Cygnus, the Swan. It is the second of these perplexing, high-speed pulsars to be found"[83] or "Researchers at the University of California at Los Angeles said yesterday they have evidence that may link chronic schizophrenia to a structural abnormality in the brain."[60] Especially near its beginning, such a story is likely to dwell on the journalist's traditional "5 W's and an H"—who, what, where, when, why, and how.

However, the news article need not be limited to that. Often what is new serves as a device for introducing information on related topics— or, in journalists' lingo, a news peg from which to hang a story. In fact, science writers sometimes prepare lengthy articles on areas of research and save them until a new development provides a sturdy peg.[72] By adding a newly written introduction, or lead, to the waiting article, the writer thus can quickly publish a piece that has news value but presents the background, too.

Sometimes not newness but another element—say, human interest, wonder, or adventure—gives a topic in science its main popular appeal. In such cases, a *feature story*, rather than a news article, is likely to result. Features, which often are longer than news stories, tend to be among the best sorts of items for portraying the process of science.

Both in print and in the broadcast media, other sorts of stories also can deal with or include science. Categories include *interpretive stories* (for example, analyses of trends), *editorials, investigative reports, columns* of various types, and *background stories* (including the sort of article that runs next to the main one and thus is termed a sidebar). Alertness to the various types of stories can aid in presenting science through the mass media.

Gathering Information

From the search for story ideas through the final revision of a piece, journalists may gather information from a wide range of human, written, and other sources. Just which ones are used, and how much, depend in part on the journalist's background, time, and place. The general reporter banging out a news story for the next edition of a small-town daily may be able only to call up a scientist or two. On the other hand, notes production assistant Barbara Costa, the research for a single episode of *Nova* can occupy a team for two months and then continue during filming and editing.

In science writing as in other journalism, interviews are a mainstay of information gathering. Who gets interviewed? Those scientists whom journalists can identify, and especially those whom they know as good sources: easy to reach, cooperative, likely to provide the sorts of information needed for a popular article, and, if possible, full of quotable quotes. Reporters specializing in science know many scientists to interview, and they can readily identify others through various routes. The general reporter, however resourceful, may have much less idea of where to turn.

Among the sources most identifiable, as well as most giving of what reporters want, are the Carl Sagans and Margaret Meads, a group termed the "visible scientists."[38] Accessible, articulate, and colorful, these are sources of whom even the most science-shy journalist is likely to know, and who most often deal with the press. Science administrators receive the next most play per person.[87] For example, popular articles on marijuana have most frequently cited not those scientists whose papers on the topic have been most widely cited by their peers but rather such individuals as the heads of health-related institutions.[81] Finally come the bulk of scientists, ranging from those whose comments in their own fields appear fairly regularly in the press to the majority, who, as science-communication scholar Rae Goodell puts it, "are rarely heard from, not necessarily because they are sullen and hostile, but more likely because they're uninterested or uninteresting."[87]

Being interested and interesting is important in ensuring that sound information on science reaches the public. Scientists can make themselves accessible and helpful in various ways. For example, they can contact the Scientists' Institute for Public Information (address: 355 Lexington Avenue, New York, NY 10017; phone number: (800) 223-1730 or (212) 661-9110) and offer to be listed by its Media Resource Service, which helps journalists to find appropriate experts to interview. They can tell the public information offices of their institutions and professional societies that they are willing to deal with the press. They can be alert to journalists' constraints. They can follow the principles of presenting science to the public, the guidelines for interviews, and other advice on working with the mass media. Then, quite likely, journalists will consider them both interested and interesting.

Preparing the Story

Once the information has been gathered, and sometimes beginning before it is all in, the journalist prepares the story. Here, as in choosing story ideas, there are time and space constraints. The reporter covering an event for a newspaper, a wire service, or the television or radio news may have only minutes in which to write; thus, little chance exists for revision and polishing. The staffs of magazines and their broadcast equivalents generally have more time to prepare and check their work, but their deadlines can still be tight. Even those with months to produce a book or documentary can have a high ratio of work to time and thus be rushed.

As for space, often the reporter may be permitted only a few hundred words in print, or even fewer over the air. Even the feature writer, book author, or documentary producer usually has much more information than can be used. Thus, the journalist must choose what to include and what to exclude, and therefore what to emphasize. The decisions, such as the angle from which a reporter chooses to write the story, are not always those with which scientists agree.

A related choice is how to structure the piece. "Write in a way that can be cut," says science writer Robert Cooke, "That's a fact of life on newspapers." Thus, in the traditional format for a news article, the main point is summarized first, in the lead; details of decreasing importance follow. Figure 8 shows an article that largely follows this arrangement,

FIGURE 8. *"Scientists Say Trees May Communicate with Chemicals" by Warren E. Leary, AP Science writer. Reprinted with permission of the Associated Press.*

WASHINGTON (AP)—You may not be able to hear it or smell it. But hidden in the rustling of the leaves may be the odors of trees talking.

Scientists said Sunday they have preliminary evidence suggesting that trees may communicate with one another through airborne chemicals when they are under attack by insects.

If confirmed, the findings from the University of Washington in Seattle would mark the first time plants have been shown to emit chemicals that convey information to others and trigger responses, said the National Science Foundation, which sponsors the work.

Drs. David F. Rhoades and Gordon H. Orians said their findings could have far-reaching implications in pest control programs when coupled with recent discoveries about the defensive systems of trees.

Dr. Jack C. Schultz of Dartmouth College announced last year that he and colleagues in New Hampshire found that several varieties of trees change the chemical composition of their leaves to ward off insects and disease.

When under attack, Schultz said, sugar maples and oaks downgrade the nutritional content of their leaves or raise the levels of toxins to discourage pests.

Orians said in a telephone interview that these recent developments are forcing scientists to re-examine their opinions about trees, long underrated in terms of complexity and sophisticated behavior.

"It shows you don't have to have brains to be clever," Orians said.

The researchers believe trees may produce and disperse chemicals called pheromones to get their warnings to neighbors. Pheromones are well known in the insect world as sex-attractants and attack stimulators, but have not been noted in plants.

The Washington scientists, in a report to the foundation, said field experiments by Rhoades found evidence that leaf damage to Sitka willow trees by western tent caterpillars and fall webworms led to changes in the nutritional quality of their leaves.

But, to the researchers' surprise, leaf quality also declined in trees up to 200 feet away that were not being assaulted by insects.

"This effect may be due to a defensive response in unattacked trees stimulated by volatile compounds emitted from attacked trees," they said.

Orians said similar reactions have been seen in red alder trees as well, "but not as strong as with the willow."

"Presumably, this is not a reaction unique to one or a few species, but we have not yet had the opportunity to examine others," he said.

The researchers have received a new grant from the foundation to experiment with Sitka willows confined in chambers to see if they can isolate and identify chemicals that might be responsible for the trees' behavior.

"If the general hypothesis is correct," said the scientists, "it is necessary that damaged plants emit volatiles which differ quantitatively or qualitatively from those emitted by undamaged plants."

which in its strictest form is sometimes called the *inverted pyramid.* Such a structure allows editors to reduce the story easily to the length desired: any number of paragraphs can be chopped from the end, and the article still should be comprehensible. Likewise, the reader need not finish the item to get the gist of it.

Despite its advantages in reporting robberies, elections, or football games, the inverted pyramid often is not the ideal format for portraying science. Unless the main point is one that the public can readily appreciate, the approach simply may not work. Thus, even if a science item is more news than feature, the journalist may choose a structure that attracts the reader with material already familiar and of interest, use that as a bridge to background information, and gradually build toward the central idea.

Whatever its format, the story must begin. Structuring a lead that draws interest and compels the audience to read on—yet is not inaccurate or trite—can be among the greatest challenges for the journalist. Once the lead is in place, the hardest part may be done. Still, however, the reporter must prepare a story that will be not only clear and interesting to the public but also acceptable to the editor.

And That's Not the End

Information gathering and writing are not the only tasks in preparing a story. Visuals may be included along the way—sometimes as an integral component, sometimes as almost an afterthought. At some magazines, articles undergo extensive checking for accuracy. Perhaps most crucial, the journalist surrenders the story to the editor, who typically lacks a background in science; unlike scientists submitting journal articles, reporters often cannot control what happens to a piece once it leaves their hands. The editing can range from barely polishing language to radically altering content, slant, and style. Also, editors decide where in a publication to place an item (and whether to include it at all); they thus influence how much attention it receives.

For many a piece of journalism, the end of a project is its beginning: the newspaper headline, the blurb on a magazine cover, or the anchor-person's lead-in is usually last to be composed. In most cases, this portion is written by someone other than the reporter who prepared the story. Thus, not surprisingly, it often is where the most distortion occurs. When you see the headline of a science story, you (and the author of the piece) may well cringe (or worse). But read on. Almost certainly there are better things to come.

This chapter has outlined how journalists cover science. Struggling against "the three tyrannies—deadlines, headlines, and space,"[49] as well as against special constraints imposed by the subject matter of science, reporters often cannot work in ways that scientists find most convenient or convey information as accurately and completely as scientists might wish. Nevertheless, what they present can greatly enhance the public's understanding of science, especially if scientists know how the press operates and thus can cooperate effectively with it.

Chapter 4

Facing the Problems

Science writer Victor Cohn of *The Washington Post* has observed that "Scientists are to journalists what rats are to scientists."[41] On the other hand, more than one scientist, speaking much less jocularly than Cohn, has conferred the label of rat on a journalist. Presenting science in the mass media can pose difficulties. Recognizing some of their origins will aid in dealing with those problems that can be avoided or combated and in accepting those problems that are inevitable.

Defining the Role of the Media

What is the function of the press? Many answers have been given, and many contain elements of validity. When a journalist and a news source define the role of the media differently, a clash can be in store.

For example, a scientist may consider television a way to educate the public about science. But the broadcaster may be out to entertain. Or the scientist may view the media as a means to promote science. But the journalist may well agree with science writer David Perlman that "We are in business to report on the activities of the house of science, not to protect it."[63] A reporter may even prefer somewhat of a "watchdog" stance.

The recognition that scientists and journalists may have different reasons for presenting science to the public can help both parties to work smoothly together. It can also aid scientists in developing realistic expectations about science coverage.

Dealing with Uncertainties

Science is full of uncertainties, and the application of science may contain even more. Findings can be tentative, mechanisms incompletely understood, effects only partially fathomed. And just what they will mean for health or the environment or the stock market is yet more unsure.

Such uncertainty makes science difficult to convey through the mass media. A clear and tidy story is easier for both scientists and journalists to tell than one that is full of *perhaps*'s and *may*'s and *if*'s and *we don't know*'s. Also, just how uncertain we are can be especially difficult to portray without statistical terminology, which usually the journalist cannot translate and the public cannot comprehend.

Yet, for various reasons, depicting the gaps and uncertainties is crucial in presenting science to the public. First, realizing what is unknown is important both in making personal decisions and in formulating public

© 1980 Sidney Harris
American Scientist magazine

policy. Second, only by including the uncertainties can we convey a realistic picture of science and foster reasonable expectations of what it can achieve.

Third, the blurs and the blanks, as well as the struggles to resolve them, make the picture much more fascinating. Hearing a final score cannot replace watching a suspenseful game. Election returns seem dull if you have not followed the political campaign. Likewise, as science journalist Boyce Rensberger observes, "Science writing that waited for the final bit of confirming data would be about as interesting as a mystery novel that skipped the work of the detectives and told you right off the bat who done it."[72]

Deciding When to Release Findings

Both between and within the scientific and journalistic communities, just when to release findings to the public is far from a settled point. One common view among scientists is that the investigator's peers should review information before it becomes available to the press. In the case of *The New England Journal of Medicine*, a policy on this issue is formalized as the Ingelfinger Rule, which is named after a former editor of the *Journal*. The *Journal*, potential authors are warned, may refuse to publish a report because the content has appeared first in the popular press or elsewhere. This policy, it is argued, helps ensure that the results of research are adequately reviewed before they reach the public and the profession, that patients do not learn of findings before their doctors see the data, and that the content of the *Journal* retains its newsworthiness.[70, 89]

Sometimes, though, pressure exists not to wait for publication or even for thorough review. Journalists may want to know about research now. Public health or welfare may rest on disseminating findings fast; here even the Ingelfinger Rule can be successfully bent.[70] Policy decisions sometimes call for whatever knowledge is available, incomplete as it may be. Also, the worthy task of portraying the process and uncertainties of science often means presenting findings in unfinished form.

Deciding when the press should receive access to results can be difficult. Little guidance can be offered other than to consider how a choice might affect the public's interests and one's own. Should journalists seek information before you are ready for publicity, you may want to note their interest and contact them once you are willing to disclose your results.

Discussing Implications and Applications

Implications and applications often are blatantly newsworthy; the rest of science frequently is not. This is another area in which scientists' and journalists' perspectives diverge.

Caution is a byword in science. Overstepping data is unacceptable in interpreting data for peers; speculating for the public is construed as even worse. Yet often the practical implications, even if not fully known, are what interest reporters and the public the most. Yes, you have found an intriguing correlation between this foodstuff and that disease—but should we change our diet or not? Sure, your analysis reveals some interesting trends—but what does that mean for economic policy? Right, you have devised a clever process—but will it help the country save energy, and if so when? As this chapter's discussion of uncertainty suggests, such questions are especially difficult to answer. Yet, unless scientists discuss the implications, it is likely that those less qualified, and perhaps less restrained, will.

Like implications, applications are only a small part of science, but they especially appeal to the media. "Journalists are under pressure to come up with 'news'" such as new medical treatments, new inventions, and new techniques to boost productivity, explains David R. Lampe, editor of *The MIT Report*. "Although this may be changing, they have spotlighted these results to the exclusion of all the interesting work behind them. They like to take the fruit and throw the tree away."

Defining Accuracy

"Whereas science is accurate to ten decimal points," Arthur J. Snider, science editor of the *Chicago Daily News*, observed, "newspapers like to settle for round figures."[53] Thus, although both scientists and reporters strive for accuracy, they define the concept differently. The journalist may consider accurate a story that captures the gist of a scientist's message and conveys it in a way that a general reader will grasp. The scientist may call for precision and completeness instead.

The precision that the scientist values may succumb in what has been called "the dilemma of comprehensible inaccuracy or incomprehensible accuracy."[53] Communicating with the public often entails simplifying terms, omitting details, and skimming over exceptions. "If you put in every possible exception and each little twist, people won't read it," says science writer Mitchel Resnick, a former *Business Week* correspondent. "And it's important for people to read about science." The skillful science writer usually can make science understandable to the public without grossly distorting the main ideas. Almost invariably, however, some of the subtleties are lost.

Not surprisingly, then, scientists consider omissions among the most common, if not the most common, errors in the popular press. In one study, clippings of newspaper science stories were sent to the scientists cited, along with a questionnaire regarding the sorts of errors found. Four of the five most frequently mentioned types of errors, each noted in about one-third of articles, pertained to omission. "Relevant information about method of study omitted" led the list; "relevant information about results omitted," "names of other investigators on research team omitted," and "qualifications of statements omitted" were not far behind. Related common objections included "story too brief" and "continuity with earlier work ignored."[84]

"Scientists have to realize that all the nitty gritty isn't going to be reported," emphasizes science reporter Karen Birchard of the Canadian Broadcasting Corporation. "They wouldn't want to read five pages in the newspaper about the ins and outs of the city council meeting."

Assessing stories one by one may make coverage seem less extensive than it is. For instance, although an individual newspaper article may seem fragmentary, the newspaper's continuing coverage of the topic (and thus the regular reader's exposure) may be much more complete. Also, even a brief mention in the media may increase the audience's receptivity to further information on a subject.

Another issue regarding accuracy is that of reliability versus validity. Both the scientist and the journalist want to get both the facts and the perspective straight. But especially if untrained in science or inexperienced in covering it, the reporter may get the details right but the interpretation at least slightly wrong. Likewise, although various magazines employ fact checkers to confirm that the bits of data in an article conform to those in source materials, the validity of those bits and of the bigger picture that they create may be assessed less carefully.

One measure that helps to make reporting in the mass media more accurate, at least in scientists' terms, is the scientists' reviewing stories before publication. For example, in the study noted above, scientists found a mean of 3.50 kinds of errors in the articles that they had read before publication, versus 6.69 in those that they had not. Scientists and journalists often disagree, however, about the value of such review.

Determining the Final Say

On the one hand, some scientists refuse to be interviewed unless granted total control over what journalists produce. On the other, some

Momma

By Mell Lazarus

MOMMA by Mell Lazarus
Courtesy of Mell Lazarus and
Field Newspaper Syndicate

reporters refuse ever to let sources review their work in any way. Determining the final say can be a major source of friction between scientists and journalists.

Fortunately, intermediate positions exist that are far more compatible and perhaps more common. For instance, the scientist may offer to review a section of the reporting for technical accuracy. Some journalists will ask sources to do so, especially if the topic is unfamiliar. Unless agreed otherwise the scientist should not suggest changes in style.

Time constraints can keep the journalist from checking back with a source. Sometimes, for example, a reporter must conduct an interview and then submit a story minutes later. Even if an article is being prepared for a weekly or monthly magazine, sending material to the scientist may still take too long. One compromise is for the journalist to read the relevant passages to the source by telephone: comments thus can be obtained immediately and possible changes discussed.

Such collaboration can be in the interest of everyone: the scientist, the journalist, and the public. Except in the very rare instances that it is agreed otherwise, however, the journalist does have final say. Or, to return to Victor Cohn's analogy, "Scientists are to journalists what rats are to scientists."

Chapter 5

Understanding Why
Journalists Do Interviews

The interview is as important an information source to the journalist as the physical examination is to the clinician or the field trip is to the naturalist. It also is almost surely the channel through which scientists can most influence the presentation of science in the mass media.

Many scientists have opportunities to present science to the public through this route. Although a scientist's likelihood (the unenlightened might say risk) of being contacted by reporters depends on such factors as field and institution, it often is considerable. For instance, in a sample consisting of 111 scientists at Ohio State University and Ohio University, 67.6 percent stated that they had been interviewed by a journalist at least once; the median number of interviews per scientist had been 4.6.[28]

Some scientists hesitate to deal with the press because they have had a bad experience or heard of one. However, as Rae Goodell of MIT points out, few scientists would abandon research simply because one experiment failed. "In most cases," she notes, "when you have had a bad experience you learn from it and know how to handle that situation better the next time."[40]

Of course, the hard way is not the only way to learn. This chapter and Chapters 6 and 7 are intended as an alternative, or at least as a supplement. This chapter explores why journalists do interviews, and the following two address how to prepare for an interview and what to do during it.

"I stated all of the important findings in my journal article. Why can't the reporter just read it and leave me alone?"

"That journalist is asking me questions answered in any introductory text. Can't he even find his way to the library?"

45

"The correspondent wants *me* to show her around the lab. Why can't one of my postdocs do it? Doesn't she know how valuable my time is?"

"Why do journalists do interviews? Can't they let scientists work in peace?"

Journalists have many reasons to interview scientists. "Even if the reporter has the time and experience to go to the source material," science writer Mitchel Resnick emphasizes, "oftentimes the original research reports don't give the perspective necessary." Resnick notes that journal articles tend to be written for peers, who readily can put a development into context. Journalists, however, frequently must conduct interviews in order to obtain background and recognize implications.

Sometimes, too, the reporter's schedule, location, or training precludes obtaining or understanding the written material. Or a journalist may be seeking data unavailable in print. In either instance, interviews can be the only option available.

Adding a human element to a story is another function of interviews. Portraying people, especially those in action, lends human interest. Likewise, as one journalist puts it, quotations "add color to a black-and-white article." Quotations also allow a reporter to present opinion with authority, as well as without seeming to editorialize. A science writer may well be able to state confidently that finding the W and Z-zero particles was important; such an individual surely can find many assistant professors and adolescent physics whizzes who would issue statements to that effect. But how much greater the impact is when the journalist quotes Nobel Prize winner Sheldon Glashow as saying "Nothing as momentous has happened in particle physics in the last twenty years."[23] Interviews are especially crucial for the broadcast media; in the words of National Public Radio science correspondent Ira Flatow, "We have to have a voice."

Information obtained through interviews also can help journalists to portray the process of research. "It seems too much like magic if you keep presenting only what science has done," Resnick states. "Science appears to move so quickly when you read only the results." Depicting the challenges, Resnick notes, can help to foster realistic popular expectations of science; in particular, public awareness of the time and effort entailed may aid in obtaining funds for research.

Finally, especially once journalists have established rapport with scientists, they may contact them for other reasons. For example, they may call to scout for new story ideas, to check whether topics for potential articles are sound, to identify sources of information, to obtain informal peer review of what other scientists have said, and to ascertain that they have reported facts accurately and interpreted them correctly. Such ongoing interaction gives scientists a particularly fine chance to assist in presenting science through the mass media.

Reporters thus have many reasons to do interviews, and scientists' availability when journalists call can help ensure that science is presented well. Mere presence is not enough, however; scientists also should know how to function most effectively when they meet the press. For guidelines in this regard, read on.

Chapter 6

Preparing for the Interview

"A beard well lathered is half shaved," an Italian proverb states. This principle pertains especially well to presenting science through the mass media. Granted, stating a definition or statistic may take little preparation, and too much rehearsal can rob an interview of its spontaneity. Nevertheless, following the guidelines below can help yield effective interviews.

Anticipating Attention from Journalists

Is a major journal about to publish your findings on a consequential or controversial subject? Did you just win a prestigious award? Do other elements make you or your work newsworthy? If so, be prepared for possible calls from the press. "Try to clear your calendar," science writer Warren E. Leary of the Associated Press advises. "There's nothing more frustrating than to call up and find that the scientist is in Bolivia." If you will not be available, arrange to have someone else cover for you. Even think about how you look; few scientists wish to make their television debuts when wearing a stained T-shirt or needing to wash their hair.

Returning Calls Promptly

As Carol L. Rogers, who heads the Office of Communications and Membership at the American Association for the Advancement of Science, notes, "Calling the reporter back the next day is often tantamount to not returning the call." Tomorrow, or even after lunch, may be too late. When leaving a message, the skillful and considerate reporter will specify the latest that you should return the call. If secretaries or others usually answer your phone, emphasize the importance of their obtaining this information and of tracking you down immediately if the deadline is tight. If you do not know the reporter's timetable, return the call right away. If you do know it, try to stay well within the constraints given; the more time the journalist has, the better a story is likely to result.

Ascertaining the Reporter's Identity and Plans

"Identifying the publication or station represented by the reporter is an obvious but often ignored practice," Martin S. Bander, director of news and public affairs at the Massachusetts General Hospital, observes in his advice-packed article "The Scientist and the News Media."[5] Not a few scientists have discovered the importance of this practice once their statements to friendly sounding freelancers appeared out of context in sensationalistic tabloids. If those requesting interviews are unfamiliar or anything seems suspicious, consider checking their credentials (for example, by calling your institution's public information office or the place where they claim to work).

In a more positive respect, knowing the nature of a respectable publication or program for which you are being interviewed aids in preparing helpful responses. Consider asking journalists for samples of their previous products, copies of the publications for which they are working, or both. Especially if you have not encountered the publication or broadcast before, ask about its audience.

"It's a good idea for the scientist to interview the journalist a little," says Ellen Ruppel Shell, a senior editor at *Technology Illustrated*. Reputable journalists rarely object to questions about their background. Obtaining the answers can facilitate the interview for both parties. Thus, if you do not know the reporter's background, find it out. Is the interviewer a general reporter, a science writer covering many fields, or perhaps someone focusing on your own specialty? How much does he or she know about the subject at hand? The first question may best be asked directly; the answer to the second often is obtained most diplomatically by conversing a bit about what the journalist hopes to ask and do.

Yes, what the journalist hopes to do. Is the story a firm assignment or a freelance piece being written on speculation? Just what is the subject? Of what sort will the piece be, and how long? What is the deadline? If not readily apparent, why are you being approached? Answers to such questions can help you decide how to proceed (and occasionally warn you not to proceed) with the interview.

Triangulating with Others

Not only can a public information office help you to check journalists' credentials. It also can aid otherwise in preparing to meet the press. For example, it can brief you on how to deal with the various media, help in preparing effective popular explanations, and even put you through a dry run of an interview. Such offices also can alert scientists to the political context of stories for which they will be interviewed, explains Jane Shure, public affairs officer at the National Institute on Aging, National Institutes of Health.

Thus, consider contacting your institution's public information office—or, ideally, one of its staff members whom you already know and trust. (Some institutions, in fact, require that all press calls, or all those that are potentially sensitive or deal with policy, be reported to and handled through such a unit.) Other possible sources of help include public information personnel at professional societies, faculty members teaching journalism or science communication, and peers experienced in dealing with the mass media. Also, if the newsworthy item is a collaborative project or otherwise involves colleagues, coordinating your and others' contact with reporters can facilitate working with the press.

Deciding Whether to Be Interviewed

"Unless reputable scientists supply accurate information to the popular media," notes neuroscientist Neal E. Miller in *The Scientist's Responsibility for Public Information*,[59] an excellent booklet on working with the media, "the public is left at the mercy of the charlatans, the sensationmongers, and . . . the anti-intellectuals." Nevertheless, a scientist need not accept every request for an interview by a responsible journalist.

For example, general reporters sometimes fail to distinguish between related subspecialties of science. In such instances, directing the journalist to an appropriate scientist (if possible, one who not only is in the correct field but also is adept at presenting science in lay terms) is the logical approach.

Sometimes, too, information is not ready to be shared with the public. When findings are preliminary or have not yet undergone peer review, scientists may validly refuse to discuss them with the press. If you decline an interview under such circumstances, the best approach usually is to tell the journalist the reason; also, you may wish to consider promising to contact the journalist once the work is ripe for release.

Arranging the Time and Place

Shortly before the space shuttle's first flight, Karen Birchard, science and technology reporter for the Canadian Broadcasting Corporation's National Radio News, was interviewing an aeronautical engineer. The engineer was struggling unsuccessfully to convey to her how the shuttle would land. Then he came up with an idea. "You're coming gliding with me this weekend," he said. "We'll duplicate the shuttle's angle of approach as we come down. Be sure to bring a tape recorder along."

"It worked," Birchard recalls, noting that she obtained excellent material. "It really did work well."

Interviews can occur face-to-face in a wide range of settings and also by telephone. The phone is often the only option for journalists pressed for time or far away. It also can be the most efficient medium for brief inteviews; there is little sense in a reporter's traveling across town to ask the name of an expert or to check a simple definition or fact.

In general, however, interviews are best done in person—whether in an office, in the field, or in the air. Meeting with a journalist lets you present information other than verbally; you can use the blackboard, show your laboratory or other worksite, and perhaps demonstrate what you found and how. Likewise, the reporter can observe details (ranging from the charts on your walls to the expression on your face) that elicit good questions and contribute to a full and interesting story. Holding the interview in person also allows you to introduce reporters to those with whom you work and thus helps show that science is collaborative. It facilitates giving the reporter written materials and visuals. It allows more feedback than is possible by phone; you can more easily check that the journalist has the story right, and the journalist has more cues from you. And it aids in establishing rapport. In many ways, a personal interview thus helps foster thorough and accurate coverage.

In addition to deciding on a mutually acceptable place, ask the journalist how much time is likely to be necessary. Set aside at least that long, and arrange to keep interruptions to a minimum. "You're not doing a journalist a favor to let him hear your telephone conversations," one science writer emphasizes.

One question is whether anyone else should join you for the interview. Although a public information officer's presence can sometimes aid at presenting information and is mandatory at some institutions, it tends to stifle conversation with the journalist. However, having someone on call to help with tasks such as photocopying is sometimes worthwhile. Also, especially in the broadcast media, a joint interview with a colleague may yield a lively, informative piece.

Providing Written Materials

Written items are rarely a substitute for an interview, but they can be an excellent supplement. They aid in preparing an accurate story; the reporter is less likely to misstate a statistic, muddle an explanation, or omit an important finding if such information is in black and white. And if provided beforehand, they help the journalist to conduct the interview efficiently. Just what to provide depends, of course, on the journalist's background and plans. But in general, as science communication expert Sharon Dunwoody notes, "One excellent strategy is to provide relevant research papers *and* a press release that details your research in plain English."[26]

Preparing Oneself

A scientist may be able to define a term, identify an appropriate expert, or summarize a process on the spot. When, however, the discussion will be more extensive, preparing oneself for the interview can be wise.

"Don't hesitate to ask for 15 minutes or so before calling the reporter back," a booklet from the American Chemical Society states.[16] Rarely is a journalist's deadline so tight that a source cannot have at least a few minutes to prepare. Taking time to construct informative, non-technical responses can speed the reporter's job in the long run and help yield a complete and accurate story. So, if you feel that you could not give your best answers immediately, request some time.

Asking the journalist what topics the interview is likely to cover often proves worthwhile. Based on what you hear and infer, you can think about what to say in the interview and how to phrase it best; preparing notes can be of help. In addition, if time permits, become familiar with the broadcast or publication for which you will be interviewed.

The journalist may be looking for sights, sounds, or both to illustrate the story. Consider asking if this is so. Then think ahead. What background sounds might be effective in a radio piece on your work? What action shots can you suggest for a television documentary? What photographs, diagrams, or graphs could clarify and enliven a written piece? If you have actual materials, as well as suggestions, to offer, be ready to provide copies of them.

Finally, consider additional sources to suggest. Especially if the story is not solely about your work, think of other experts to recommend. If it will discuss a controversy, perhaps be ready to suggest speakers on each side. In addition, try to identify reading materials and other items (for instance, lectures, exhibits, and tapes) that may help the reporter produce an informative, interesting, accurate, balanced piece.

This chapter has discussed preparing for the interviews. The beard is now half shaved; onward to the other half.

Chapter 7

Making the Most of the Interview

In publicizing as in pursuing science, preparation accomplishes little by itself. One must also follow the proper rules, apply the proper skills, and remain alert for pitfalls to avoid and opportunities to pursue. Thus, this chapter offers guidelines for making the most of the interview.

Laying Down Ground Rules

Many years ago, a luncheon for science writers featured as its speaker a prominent scientist, then serving as a consultant to the government. Much to the journalists' delight, the scientist candidly criticized the administration for which he worked. Then he noted that his remarks had been off the record, "of course." His statement, however, came too late. And everyone present filed a story.

If you will be saying anything that you do not want to see in print or to have attributed to you, say so before the words leave your mouth; such is the rule in journalism. Be sure, too, that the reporter agrees to the restrictions, for otherwise you proceed at your own risk. In other words, if anything is *off the record* or *not for attribution*, say so ahead of time; and consider that you may be better off not conveying the confidential information, or making the unprintable statement, at all.

Be specific about how much attribution, if any, is acceptable to you. Although journalistic jargon exists for these levels, saying exactly what you mean is probably the most reliable approach. For example, you may wish to state, "This information is only to help you understand the situation; do not put any of it in your article" or "If you want to quote what I'll say now, you can attribute it to 'another physician who has treated many patients with such injuries,' but don't get any more specific than that."

If you want to lay down any other sorts of ground rules, also do so as

54

early as you can. At the beginning, before you provide the information that reporters want, you have the most leverage. Thus, for example, if you insist on having all quotations read back to you or on checking the description of your work for technical accuracy, strike an agreement right away.

Avoiding Embarrassment

In the excellent article "Between Scientists and Public: Communicating Psychological Research Through the Mass Media,"[56] Robert B. McCall and S. Holly Stocking write: "It is a good policy for you to imagine that you are always talking to 10,000 or more people throughout the entire interview—sometimes you are." Anticipating how a comment would look in print or sound over the air can save embarrassment. So can declining a question (and, ideally, suggesting a qualified source), if you lack the expertise or the authority to answer it. Most important, so can following economist Murray L. Weidenbaum's first three rules for dealing with the press: "Do not utter falsehoods," "Avoid prevaricating," and "Never lie."[65]

Taking Notes

Although your tape recording an interview may motivate the occasional careless (or worse) reporter to be more accurate, it rarely is worth the alienation it can cause. Keeping pad and pencil handy, however, is rarely offensive and often useful. Many of us find that jotting down ideas and phrasings helps us to express ourselves. In addition, briefly noting major questions and replies can aid in learning to present science through the media, especially if the notes are compared with the story that ensues.

Presenting the Information

Following a few basic guidelines for presenting the information both aids journalists and helps ensure effective coverage. "The best advice in interviews is to *get to the point and stick to it*," emphasizes editor David R. Lampe. Reporters, public information specialists, and media-wise scientists agree that concise, well-focused answers work best. Sources who talk around their topic, ramble from idea to idea, and burden interviewers with copious detail not only complicate the work of journalists. They also invite extensive editing, which can inadvertently or otherwise distort their material.

"The popular news media channel is very noisy," Robert M. Byers of the MIT news office adds. "Therefore, repeat your signal several times during the interview." When a point is crucial, consider explicitly saying so or otherwise underscoring it. Also, if you have information that you consider important but that the reporter does not request (for instance, relevant anecdotes that can help the public to understand the process of science), try to present it nevertheless.

Another key principle is to *do as much as possible of the translating yourself.* Of course, doing so is crucial in interviews broadcast live. It also is important in other interviews, for the more of the material the scientist puts in clear common language, the less of it journalists must change and the less chance exists for error to occur. Thus, use simple words, relate the unfamiliar to the familiar ("Scientists who make good analogies are worth their weight in gold," one journalist remarks), include examples, make the relationships between ideas clear, and follow other advice from Chapter 1. Two models of presenting science effectively in an interview, both taken from the *MacNeil-Lehrer Report*, are the following:

> MACNEIL: Dr. Stone, first of all, what truly historic discoveries have been made this week?
>
> EDWARD STONE: Well, I think one must list the fact that we have determined that Saturn's large moon Titan has an atmosphere probably two to three times as dense as the atmosphere here on Earth—an atmosphere predominantly nitrogen, but with a trace of methane. Then, we have imaged a new class of satellites smaller than our own moon, for instance, which seem to be essentially pure water ice—very little rock. And third, we have now improved our resolution of the rings by a least six thousand times, from the earth-based images, and what we find is a remarkable system which was totally beyond our comprehension prior to Voyager.[78]

> MACNEIL: Why is diagnosis so difficult in this disease [Alzheimer's disease]?
>
> MIRIAM ARONSON: Diagnosis is so difficult because the onset of this disease is very insidious. What happens is you get a symptom here, a symptom there. You kind of can't put your finger on it at the beginning. Somebody begins to perform poorly. You get an accountant who suddenly can't add figures quite the way he used to ... Or you get the executive who starts to miss meetings, miss deadlines, make poor decisions, and people assume, well, it's exhaustion....[52]

In presenting information as you would directly to the public, consider keeping a specific person or group in mind. Patrick Young, science writer for the Newhouse News Service, sometimes tells scientists to pretend that he's a "one-man Rotary Club."[50] Similarly, a guidebook from the Los Alamos Scientific Laboratory advises scientists dealing with the media to "communicate as if you were giving a first talk on your subject to a high school science class."[53]

Giving the journalist written materials and visuals, if not provided already, also contributes to accuracy. And a seemingly apparent but often neglected aid is merely to talk slowly enough. Even if a reporter is tape recording an interview, beware of speaking too rapidly or presenting successive ideas at too great a clip; the interviewer should have time not only to record what you say but also to assimilate it and formulate further questions accordingly.

Implications are an aspect of science that scientists especially hesitate to discuss outside their own ranks. Indeed, "The Fairly Concise *New Scientist* Magazine Dictionary" defines aphasia as "loss of speech in social scientists when asked during conversations at parties, 'But what *use* is your research?'"[31] Yet the implications often interest journalists and the public the most. Regardless of whether scientists present them, reporters are likely to discuss them in their stories. Even if reporters do not, the public is likely to speculate on them. To help ensure that the reporter's discussion and the public's speculation are valid, scientists generally should present the implications themselves if asked and perhaps even if not asked. Sometimes, reviewing a grant application can aid in deciding what to say.

Especially when discussing implications, limitations and uncertainties are important to convey. Yet doing so in an interview can be a tricky art. Lumping all the limitations in one part of the discussion and making sweeping statements in another virtually invites journalists to take material out of context. Even segregating the qualifiers in one part of a paragraph or sentence can put a speaker at risk. The best tactic often is to word statements so that qualifiers are difficult to remove. For example, embed brief qualifying words and phrases snugly in the relevant sentences, and use words with the qualifiers built in ("promising," not "exciting"; "suggests," not "proves"; "treatment," not "cure").

Finally, as Byers notes, "It's not too bad an idea to find some way to say to the reporter before he or she leaves or hangs up, 'What do you think I said?'" Having the reporter read back or paraphrase the material allows you to identify and correct inaccuracies. It also gives reporters a graceful way to obtain information that they missed the first (or second or third) time around and may hesitate to request.

Retargeting Questions

Retargeting questions is a useful skill in dealing with the media. Not only can it aid in correcting misconceptions that journalists and the public hold. It also allows sources to present information that they consider important to convey but for which journalists may not explicitly ask.

Reporters, particularly those not specializing in science, sometimes ask questions based on incorrect premises. So do community members contacting call-in shows. As an example of how such a misdirected inquiry can be retargeted with tact yet clarity, note how astronomer Bradford Smith replied when asked why Saturn has rings but other planets do not:

> Well, actually there are two other planets that have rings—both Jupiter and Uranus. At one time, we thought that Saturn was the only planet that had rings at the present. Now, all planets may have had rings at one time, but rings are ephemeral things. They tend to want to go away; there are disruptive forces which make them dissipate and disappear. The real question is why, after four and a half billion years, does Saturn, Jupiter and Uranus still have rings.[78]

When questions are vaguely aimed, alert sources sometimes can direct them toward the targets that they desire, as Donald Kennedy, then Commissioner of Food and Drugs, did on *Meet the Press*. When asked "What do you consider the single-most important, most serious problem facing us in this area of environmental cancer?" Kennedy replied:

> I think it is persuading the American people to believe that this is a complex disease with a complex causation, and it is one that they share the responsibility for preventing. I think that returns us to the question of the war on cancer and whether it is the Vietnam War analogy. I think that metaphor was intended to suggest that this is one of those problems that gets more complicated the more resources you throw at it, and that we are not going to get a technological fix that allows us to forget about cancer. It is a complex disease. It has multiple origins. A lot of them are environmental. That means that everybody, regulatory agencies, citizens, everybody, is going to have to work at it on a number of fronts at once, and I think gaining some public understanding of that complexity is the most important part of the problem.[30]

An occasional bit of deft retargeting can aid in making the most of an interview.

Offering Further Help

The end of the interview is a fine time to volunteer further assistance and thus to help ensure effective coverage of science. Perhaps, for example, offer (or repeat your offer) to check material for technical accuracy. Likewise, you may well note your willingness to help in other ways, such as by answering questions that later may arise. Be sure that the journalist knows where and when you can be reached.

After talking with you, the reporter most likely will know more than before not only about the main topic discussed but also about your general field. In addition, you probably will be more aware of the sorts of information that the journalist seeks. Thus, many reporters consider the end of an interview a fine time to ask sources to suggest other topics worth covering. If you seem to have established especially good rapport, perhaps volunteer some story ideas even if you are not asked.

Even after the interview, the opportunity (indeed, the responsibility) to present science to the public does not end. If you think of additional material that could markedly improve the story or you recall an important error in what you said, consider contacting the journalist. But here, as in the interview itself, beware of bombarding the reporter with detail; a basic dictum remains "Stick to the point."

As Neal E. Miller notes in *The Scientist's Responsibility for Public Information*, "You can control the interview."[59] By knowing the rules, applying the appropriate skills, and remaining alert for opportunities, you can make fullest use of this route.

Chapter 8

Knowing How the Media Differ

"We in television have a special problem," Jules Bergman, science editor of ABC News, has observed. "We don't have a page 74 or 89. . . . Our 30-minute news show—22 minutes after commercials and promos—has only one page: the front page."[80] Both in their depth of news coverage and in other respects, the various media differ despite their parallels. Recognizing the distinctive potentials and constraints of each can aid scientists in presenting science through them.

The Print Media

Because the print media are non-scientists' main source of news about science,[24] working effectively with them is especially important in presenting science to the public. Print is typically the medium most familiar and comfortable to scientists. As previous chapters indicate, however, popular articles differ markedly from journal papers not only in their content and format but also in the way they are prepared. Newspapers and magazines likewise differ from each other in aspects useful to recognize.

Newspapers

Within each of the mass media, depth and expertise of science coverage vary widely. Among newspapers, and even within a single newspaper, the diversity tends to be especially broad. Both science writers and a large variety of other reporters cover science for newspapers. An article can be a few paragraphs written in minutes from a single source. At the other extreme are accounts like the "monster story" that *The Philadelphia Inquirer* published regarding Three Mile Island: a total of nine pages, representing the work of an estimated sixty journalists.[77]

With a few exceptions such as *The New York Times*, newspapers are largely local media. Because they tend to focus on hometown people and events, their reporters often seek local experts and local angles. By cooperating in this pursuit, as well as remaining sensitive to local concerns, scientists can help present science to the public; they also can enhance their own, and their institutions', image and effectiveness in the community. In addition, by using the locale as a common reference point ("about one percent of the population, which means over a thousand people here in New Haven," "for instance, the earthquakes that we felt in Boston last year," or "looks through the microscope rather like the traffic on the Beltway"), sources literally can make what they say hit home.

Although a local angle may help carry a science story into the newspaper, sources who also offer a broad perspective can not only aid in informing readers most fully but also contribute to nationwide science coverage. "I think that most journalists are looking for a more significant angle," says one former newspaper reporter. "A national angle on a story happening at Podunk Community College can put Podunk on the map, boosting the small town's pride and self image. It will also help get the piece onto the AP wires, something all aspiring journalists want."

Science stories in newspapers can indeed get wide play. "A scientist may think he or she is talking to a small home-town paper in Oregon," Neal E. Miller of the Society for Neuroscience notes, "and the next thing that happens is that Mother calls from Maine with congratulations about the Nobel-quality work she read in *her* local paper."[59] Wire services such as the Associated Press and the United Press International often obtain news from local newspapers, and newspaper stories are a frequent source of story ideas for journalists in various media.

Magazines

"Scientists feel that magazines do by far the best job of covering science for the interested lay public," report Sharon Dunwoody and Byron T. Scott, who conducted a survey on the subject. "And by a wide margin they prefer to deal with a magazine journalist rather than a representative of any other medium." Dunwoody and Scott propose that scientists may prefer magazines because they find them similar to journals.[28]

Other factors also may contribute to magazines' favor. Magazine articles often are longer, more slowly prepared, more reflective, and more detailed than stories in the other media. They tend to be features, which are better suited than news stories to portraying issues and processes in science. Some magazines carefully verify facts in articles. Also, many serve readerships that are highly educated, or that are especially attentive to science or otherwise specialized; thus they can cover science with more sophistication than appears in the newspaper or on the nightly news.

Groups of journalists collaborate on the stories for some magazines. At *Newsweek*, for example, the reseacher assigned to the science or medicine section, as well as correspondents in regional bureaus, gathers information; the science or medicine writer composes the stories; and afterward the researcher double-checks the facts. In this system, interviewers gather much more information than appears in print and have limited control over its use. Oftentimes stories quote or mention only some of the scientists interviewed.

Whereas some magazines are totally staff-written, many obtain articles from freelancers. Thus, merely hearing the name of a major magazine is no guarantee that the journalist's credentials are sound. Alertness to this and other aspects of how and for whom magazine articles are prepared can aid scientists in communicating to the public through magazines.

The Broadcast Media

The broadcast media, with their sounds and sights, offer opportunities unavailable in print. However, they have their own limitations as well.

In the print media, readers can scan. If they decide not to finish a story, they can skip immediately to another one; if they forget a point, they can return to it; if something is unclear, they can review it as slowly as they please. Such is not the case with the broadcast media. Audience members uninterested in an item cannot jump ahead; they must wait it out in boredom, turn it off, or go away. Thus, to hold their audiences, broadcasters concentrate on subjects and approaches of wide appeal and usually keep individual stories short. On general programs this often means little science; even on science programs, this often means little depth. Also, in the broadcast media, the audience cannot look (or listen) back. If the message is not followed while it is being presented, it simply is lost.

In practical terms, these factors mean that clarity, brevity, and liveliness are especially important in the broadcast media. Says John D. Miller, science reporter for WEWS-TV in Cleveland: "The short, succinct statement delivered with a lot of energy is what will get on TV." Thus, when you will be interviewed for a news report, be ready to present your main idea in a crisp, clear sentence or two. For any radio or television broadcast, keep your statements simple and to the point. Realize, however, that unless you are being interviewed live, your remarks, no matter how popularly phrased, almost certainly will be edited and condensed.

If you are indeed being interviewed live (for instance, on a talk show), you have greater control, but there are greater demands on you. The interviewer cannot delete qualifiers that you carefully insert, but neither can he or she erase your um's and er's or the statements that you would really rather take back. And although you are unlikely to have the journalist muddle your message (at least without your having a chance to unmuddle it), the interviewer may give you little help in translating it into popular terms. In short, how successfully you present the information is largely in your own hands.

Radio

Radio runs science stories of various lengths, from news reports lasting only seconds, to features of several minutes, to lengthier documentaries. Largely though, as Ira Flatow of National Public Radio notes, radio is "only a headline service."

Whereas all of the other media can use visuals, radio must convey its entire message through sound alone. Yet the radio, like a tape recorder, tends to rob voices of their vitality. Therefore, those experienced with radio advise, be careful to put expression in your voice, "Even if you overemphasize every single word," says Karen Birchard of National Radio News in Toronto, "it doesn't sound like you're really hamming it up."

Background noises also can enliven radio. "We always like to have sounds," Flatow states. "They're very useful in keeping people interested in the story." Thus, if you are being interviewed for radio, consider suggesting laboratory noises, animal calls, or other sounds that you think might enhance the piece.

Television

In television, the visual element adds new possibilities but also new constraints. The latter include those that its equipment places on journalists. Print and radio reporters can telephone almost anywhere to conduct interviews, but camera crews have much less mobility. Thus, if a local station is to cover science, it may have to rely heavily on showing local scientists. Also, as a member of the *Nova* staff emphasizes, "A crew can't just pick up and come back. Aside from the planning, so much money is involved." For instance, filming for a *Nova* episode costs "well over a thousand dollars a day, " not including travel expense.

One constraint that may not be as great as many scientists imagine regards how to look. "I think that most of it is not dress but just understanding what needs to be done for TV," John D. Miller states. When being interviewed on site, scientists should generally "look the way they work," he and others say. Nevertheless, some attention to appearance can be worthwhile. Experts advise those appearing on television to wear solid colors rather than patterns (which tend to look "busy") and to avoid white (which can appear too bright on the screen).[51] In addition, they tell the interviewee to look at the interviewer, not the camera, during an interview.[59] If in doubt about what to wear or do, one should simply ask.

Something else one can ask is for an occasional retake. If you realize that you could have said something much more effectively, consider requesting another chance.

Also ask whether you should suggest, or perhaps provide, visuals. They especially those that are colorful and move, enliven the piece (and thus increase its chances of being aired and watched). They also help to make effective use of the limited time available on television.

In fact, if your work may have television potential, think ahead. "If you are taking high-quality motion pictures of your research for any reason," suggests Neal E. Miller of the Society for Neuroscience, "you should save a print with the idea that excerpts from it may possibly be useful in some television program."[59] The same principle holds for other visuals.

Television is generally the most mass of the mass media. "Working with television is like gin in a bathtub," observes Keith Mielke of *3-2-1 Contact*, a science program for children. "Half an inch doesn't look like very much, but it's a lot of gin." Newspapers can publish some articles for only part of their readership, magazines are often geared for special groups, and different radio stations cater to different constituencies. Although the Public Broadcasting Service and certain cable stations do tend to serve specific sectors, television addresses the public at large. "I would encourage scientists to talk to broadcast journalists," science reporter John D. Miller, who himself is trained as a scientist, concludes. "It's the very best way to get in touch with people who are not attentive to science."

Chapter 9

Working with a
Public Information Office

A public information office—be it of your institution, a professional society, or an organization running a meeting—can help you ensure that journalists are aware of science news and that they present it accurately and effectively. It also can aid in presenting science directly to the public. "We are supposed to give such assistance," emphasizes Carol L. Rogers of the American Association for the Advancement of Science. "It's our job, so a scientist is not asking for a favor but using a resource created for that purpose." This chapter describes what PI offices do and offers pointers on how to collaborate with them. You may find some of this material particularly useful if you yourself are ever responsible for PI activities, such as when you are hosting a guest speaker or holding a conference.

Staff and Activities

Although you may recognize a finding as newsworthy or have information that you think merits public attention, you may lack the contacts, skills, and resources to initiate and facilitate press activity. For example, you may realize that a volcano, a major new theory, or a controversy is erupting and want to alert journalists. If you know a reporter with appropriate interests, perhaps all you need to do is make a call. If not, the task can be more difficult. When contacted by strangers, including scientists they do not know, journalists can be rightly skeptical. "Is this person really qualified?" they may quite validly wonder. "And could the caller be trying to sell something?"

Likewise, if you are about to publish a paper that you think will attract attention in the press, you may wisely decide to prepare background materials and a lay summary for journalists. However, you may lack the skill to do so effectively, or you may lack the time required.

Also, you and your staff may lack the facilities to reproduce and distribute such materials conveniently.

In such situations, a public information office (also known by such names as news office, office of public affairs, public relations office, and office of communications) may be able to provide what you lack. Such units (or, more precisely, the better of them) know and are trusted by journalists. They have staff members adept at preparing press releases and related materials. They also have means to reproduce such items, as well as address lists and resources for mailings to the press.

The staff of PI offices can vary in many ways. One is size. In a small organization or one that does not stress communicating with the public, PI may be the responsibility of a single person, perhaps one who also has other roles. However, a large organization or one emphasizing public contact may have a large division devoted to PI. Within a large unit, some members may be assigned to particular subject areas (at a university, for instance, one PI specialist may concentrate on engineering, another on health sciences, and so forth); getting to know the individual assigned to your field can facilitate dealing with the press. Some members of a large PI office may specialize in particular activities, such as arranging publicity for conferences, preparing lay brochures, and dealing with media calls.

PI specialists vary widely in background and ability. Many are trained in communication and related fields; a fair number have worked as journalists. Some have at least undergraduate degrees in the sciences. Others have no formal education in either area.

PI offices, as well as their staff members, vary greatly in quality. The best are highly professional outfits devoted to and skilled at communicating through the press and other means; they have earned the respect of both journalists and scientists. The worst are sluggish or fumbling— or misguidedly loyal, loudly announcing every minor advance but barring the press if controversy threatens to arise; scientists, journalists, and the public soon pay them little heed.

The time to assess a PI office is before something newsworthy arises, not afterward. In fact, foresight is the general byword in working effectively with PI offices. "Neither science nor journalism is served when my office and similar offices around the country learn of a major study from a reporter," says Elaine Freeman of Johns Hopkins, "and we learn that the principal author is in another country for two weeks. We've all lost a great opportunity to convey something about science to the public."

PI offices can have many activities, some of which will be discussed in more detail below. As James Cornell, who is in charge of public information at the Harvard-Smithsonian Center for Astrophysics, notes, these activities fall into three main categories: working with journalists, communicating within the organization, and informing the public directly.

With regard to the mass media, PI offices typically issue press releases, conduct press conferences, facilitate press coverage of meetings and other events, direct journalists seeking interviews to appropriate sources, and advise members of the organization in dealing with the press. Some PI units distribute lists of story ideas and of experts in various fields. They may also coordinate, monitor, and control press activities; for example, at some government agencies, the PI office oversees all contacts with the media. Similarly, they may be responsible for handling press inquiries about sensitive topics and policy matters.

As for in-house activities, PI offices may, for example, prepare newsletters distributed mainly to employees, members, or alumni. Some such publications reach journalists.

Third, some PI offices issue items directly for the public. These can range from fact sheets and booklets to broadcast segments (for example, the American Association for the Advancement of Science's 90-second radio news program "Report on Science," the American Chemical Society's radio interview series "Man and Molecules," and the American Institute of Physics's two-minute television spots "Science TV News"). PI offices sometimes sponsor series of lectures by scientists. They may prepare and disseminate promotional materials. In addition, they respond to routine inquiries from the public and so relieve scientists of such work. For example, Cornell observes that "we answer a lot of basic astronomy questions here."

The work of a public information office can be sensitive and difficult. "Our problem," says Paul Lowenberg, assistant director of the PI office at the University of California, San Diego, "is that we walk in both worlds. We are the university when dealing with the media, and we are the media when we're dealing with the university."[44] Understanding what PI offices do and how they do it can help close one of these gaps and thus aid in presenting science to the public. Therefore, the rest of this chapter deals with three PI activities likely to involve scientists most: press releases, press conferences, and public information activities at meetings.

Press Releases

When you and a public information specialist agree that something should be covered (or realize that it will be covered) by the media, preparing a press release can be worthwhile. Typically, the PI specialist prepares such an account, also known as a news release; in many cases, the scientist checks it before it is distributed. The release may then reach journalists through several routes: for example, as a general mailing, in response to individual requests, at press conferences, and in press rooms at meetings.

The press release alerts journalists to subjects deserving coverage. Because it presents the facts in print, it also helps reporters to convey information accurately. Sometimes press releases are published as is (especially by small newspapers), and sometimes those designed for radio or television are presented in their original form; more often, however, journalists use information from them in preparing original stories.[56] Science writers sometimes file press releases for later use.

Press releases, like antibiotics, can be powerful tools but tend to become ineffective when used indiscriminately. Science writers rapidly develop resistance to institutions deluging them with press releases every time an employee is promoted, an abstract is published, or a grant is renewed; envelopes from such sources often bypass the letter opener and land directly in the trash. In short, press releases should be reserved for the truly newsworthy.

Figure 9 is an example of a press release. Starting from the top, several points are of note. First, a release lists at least one member of the PI office to contact for further information. It also notes a release date (and sometimes time)—that is, the earliest that journalists may publish or broadcast the story; providing such a date can be important if, for example, a scientist is publishing or presenting a paper and does not want the content publicized beforehand. The general subject also appears near the top of the release.

A well-constructed press release makes its main point right away. In other words, it presents most or all of the five W's in the first few lines and elaborates on them afterward. This arrangement has at least two advantages. First, an editior who wants to publish the press release without rewriting it can adjust it to the desired length by lopping paragraphs from the end. Second, this format aids in grabbing the attention of busy journalists. "I read dozens of releases every day," said William Stockton when he was science editor of *The New York Times*, "and my rule of thumb is that if a release can't sustain my interest through the first two paragraphs and at least tell me what's going on, I won't spend much more time with it"[87].

Press releases are non-technical both in wording and in emphasis. The words are common ones; the sentences and paragraphs are short. However, a press release is more than just an abstract in popular language; it also may provide background, so that those reading it will know why the news in it is important and where it fits in. Putting the release in lay terms both allows the release to be published or broadcast as is and makes the release useful to journalists unfamiliar with the topic being discussed.

In general, press releases are short; they are often two pages or less. They may be supplemented, however, by other items such as photographs and diagrams, background materials, biographical information, and bibliographies. Sometimes, if the release focuses on a paper being published or presented, the text of the paper is enclosed (a courtesy especially welcome to the reporter whose deadline or location precludes tracking the paper down); at the very least, copies should be readily available to journalists who call. Add several of these materials (usually including the press release) together, perhaps include others such as conference programs, and maybe put them into a folder—and the result is a "press packet" or "press kit" to distribute to journalists.

Press releases are important tools for conveying science through the mass media. You can help them to achieve their potential by alerting PI offices promptly to newsworthy items, cooperating with PI specialists in preparing press releases, and being accessible to reporters once the releases are distributed. Also, consider keeping a file of press releases and related materials about your work. You may wish to give them to journalists who contact you, and rereading press releases beforehand can aid in speaking with reporters and lay audiences.

Press Conferences

By granting more than one reporter access to you at once, a press conference, also known as a news conference or press briefing, helps you reach a large audience efficiently. Essentially an interview by a group of journalists, it is an effective way to disseminate an important message, such as a warning regarding safety or health. It is also a convenient strategy if many reporters are, or are likely to be, clamoring to interview you.

Either you, a member of a PI office, or someone else may propose holding a press conference. Of course, such an event should take place only if the news is major. Calling a press conference to issue a minor announcement wastes journalists' time and, in the long run, hampers communication through the press.

If a press conference does indeed seem worthwhile, the PI staff will schedule one. They will try to set a time that suits reporters' deadlines and is relatively free from competing news events. Public information officers in Washington, D.C., for example, attempt to schedule their press conferences on "panda days"—that is, days so quiet that the media

From the News Office   January 11, 1983
Massachusetts Institute of Technology
Cambridge, Massachusetts 02139
Telephone: (617) 253-2701   FOR IMMEDIATE RELEASE

Contact: Robert C. Di Iorio
           (Home: 1-947-3290)

LOW-CARBOHYDRATE DIETS
MAY TRIGGER CARBOHYDRATE BINGES,
M.I.T. RESEARCHERS FIND

Dieters who swear off carbohydrates for a long period may be susceptible to carbohydrate binges that can put back the lost weight once a normal diet is resumed, Massachusetts Institute of Technology researchers have found.

The scientists came to the conclusion after studying the behavior of rats that had been given protein-rich, carbohydrate-restricted diets not unlike some of the diets followed by weight-conscious Americans.

"When animals previously restricted to a diet containing only protein and fat are allowed to choose between (low-carbohydrate or high-carbohydrate) diets," the researchers write in an article to be published in the January Journal of Nutrition, "they respond by over-eating carbohydrate compared with control rats: they increase the grams of carbohydrate consumed and the proportion of their total calories represented by carbohydrate."

The researchers who conducted the study--all from MIT's Laboratory of Neuroendocrine Regulation--are Dr. Judith H. Wurtman, Peter L. Moses and Professor Richard J. Wurtman.

The scientists suggest that the carbohydrate-deprived rats cannot control their subsequent over-consumption of carbohydrates, a situation that presumably would be the same for humans. Nor is the over-consumption attributable in the rats to the need to consume more calories because, in other experiments, the rats did not overeat protein or fats, the researchers found.

"The overeating of carbohydrate in carbohydrate-deprived rats probably reflects an alteration (caused by the period of carbohydrate

(MORE)

FIGURE 9. *An example of a press release.*

WURTMAN 2-2-2   January 11, 1983

deprivation) in the brain mechanism that normally regulates carbohydrate intake," they write.

The scientists also demonstrated in another study using normal,
non-deprived rats that carbohydrate intake in one meal is adjusted to
the amount of carbohydrate consumed in the previous meal.

They found that carbohydrate consumption was diminished, in a
diet-choice situation, among rats that had eaten a small carbohydrate
pre-meal one hour earlier.

"The response of the animals to this carbohydrate pre-meal was
specific," they note. "For example, while they consumed as many total
calories as the control group, they chose to eat fewer of their
calories as carbohydrates. This affirms that the animals were
responding to the carbohydrate and not the calories ingested, and provides further evidence that distinct mechanisms regulate appetites for
total calories and for carbohydrates."

At the conclusion of their report, the scientists consider the
implications of their findings for human dieters.

"The animals' ability to regulate carbohydrate intake when
allowed to eat a carbohydrate-containing diet and their inability to
do so when given carbohydrate-free foods should be considered in
evaluating the efficiency of (low-carbohydrate) weight reducing
regimes," they comment.

"Although weight is lost rapidly on such diets, the individual's
ability to control his appetite for carbohydrates when reintroduced
may be compromised, causing over-consumption of the nutrient and
weight gain...that so often accompanies the termination of (such)
diets."

In an interview, Dr. Judith Wurtman said the study may explain,
in part at least, "why there is such rebound-eating after a low carbohydrate diet is over. If nothing else, this research should remove
the guilt from an ex-dieter who thinks his craving for cookies is
simply emotional." She likened such persons to sleep-deprived
individuals who find themselves sleeping longer over the weekend.

                            --END--

FIGURE 9. *(Continued)*

cover the pandas feeding at the National Zoo. Nevertheless, the best laid plans can go astray, as when the National Institutes of Health meticulously arranged a day of press briefings but then President Nixon resigned.

In addition to contacting reporters and obtaining a properly equipped room, the PI staff probably will arrange to distribute background materials. "Never show up at a news conference without some paper," Warren E. Leary of the Associated Press advises, adding that handouts are most helpful if reporters receive them a day or two in advance. At press conferences as in other situations, such items foster thorough, accurate coverage.

At the press conference itself, a PI specialist normally introduces the proceedings. Next, the expert or experts being featured present the main points. Of course, such presentations should use effective common language; in preparing them, a PI specialist can be of help. Because broadcast as well as print journalists may be present, statements should be not only clearly and quotably worded but also well suited to the ear. In addition, speakers should remember that television cameras may be focused on them. "Often, long after they have forgotten what you said," Martin S. Bander of the Massachusetts General Hospital notes, "the public will remember how you looked or sounded."[5].

Generally, after the prepared statement(s) come questions from journalists; now is a time to follow the basic principles of making the most of an interview. Then the PI specialist who opened the conference will bring it to a close. The burst of activity is over, but attention from reporters may not end. For instance, journalists may individually approach you immediately afterward. And especially if you have proved to be the kind of source that journalists favor, your increased visibility may well engender future calls from the media.

Public Information Activities at Meetings

Meetings, including the annual conventions of major scientific societies as well as conferences on specific topics, are an important source of science news. Before, during, and after such events, the organizations running them, and sometimes also those where the speakers work, conduct various public information activities. Knowing what is done can help scientists to cooperate in seeing that information reaches the public effectively.

The PI activities frequently begin long before the meeting itself. In fact, if a sponsoring organization's mandate includes informing the public about science, speakers and topics may be chosen in part for their news appeal. The PI staff may send the press various materials in advance. For example, a brief announcement of the meeting may be issued well beforehand, to be followed by a relatively detailed press release highlighting sessions that journalists are likely to find especially interesting.

At the meeting, the press room (or press suite) is the headquarters for media activity. Generally, a PI professional is in charge; various numbers of staff may assist. These personnel give journalists background material, help them identify newsworthy stories, and handle other logistics. Sometimes, especially at a meeting on a fairly narrow technical topic, having a scientist in the press room to answer questions can be worthwhile. If you are adept at translating science into popular language and like to deal with the press, consider volunteering for such a role.

The PI staff also arranges interviews for journalists. Therefore, speakers at a conference should let those running the press room know where and when they can be reached. They also should come prepared for possible interviews.

The press room contains more than people to provide information and services. It also has typewriters and telephones for journalists to use, coffee for them to drink, and perhaps something for them to eat. Especially important, it is likely to have lots for them to read—texts of papers being presented, press releases on the sessions, and related items. Before a conference, you may be asked to send or bring such materials. Even if not, you may be wise to have them with you, in case PI staff or reporters ask for them.

Press conferences, too, are sometimes an important part of the media activity. At a meeting dealing with many topics, they may be scattered throughout; at a meeting on a specific subject, and particularly at one culminating in recommendations, a single press conference may occur at the end. Some press conferences are arranged well ahead of time, and others suddenly materialize when unexpected issues emerge.

Press conferences, as well as the workings of the "science writing inner club," help ensure that journalists attending a meeting file stories as their editors expect, and that they choose topics soundly and cover them well. In the short term, the situation may mean that many journalists focus on the same few stories, and thus that few of the topics presented reach the popular press. In the long run, however, the effect may be quite the opposite.

The reason? Press conferences let reporters prepare stories quickly. Thus, time remains to amass ideas, materials, and contacts for future use.[50] The journalists can circulate from session to session. They can keep picking up items from the press room. They can move informally among the scientists present and so learn who is doing newsworthy work, hear who can speak well, and establish rapport. (Science writers, like scientists, realize that much of the truly new emerges not at the formal sessions but in the corridors.) Long after a conference, reporters may draw on the information obtained and the contacts renewed and made.

Even once the meeting is over, public information efforts and media activity can persist. Shortly after a conference, the PI office may issue a press release on the event. Conference summaries or proceedings issued a year or more after a meeting can still attract the press. Meanwhile, the PI office or a clipping services hired by it may monitor the media for references to the conference and might send you copies of articles in which your name appears. And long after the meeting you may get press calls generated by it—and so have opportunity to refer to this book.

Chapter 10

Succeeding with What
Succeeds Media Attention

The interview is over. The conference is done. The phone has stopped ringing after the press release. Yet demands and opportunities related to media attention may not have ended. Thus, this chapter discusses succeeding with what succeeds interaction with journalists.

Reacting to Coverage

The main item that succeeds interaction with a journalist is likely to be a story. Recall, however, that not every story makes it into print or onto the air. Other facts worth remembering include the following. A brief news item can capture only the highlights of a development; it tends to emphasize findings and implications rather than aspects such as methods. Especially for a feature story, the reporter typically gathers much more information than appears in the final product. Also, the journalist who composes an item usually does not write or check its headline or title.

Scientists who have been interviewed, or who have spoken publicly, on controversial topics should realize, too, that they may be criticized in the press or at least have contrary views presented along with theirs. Drawing on his own experience in this regard, psychiatrist Robert Dupont advises: "Understand that you are not charging in with the answer, that the water is not going to part before you, and that everyone is not going to say, 'Great. Now I understand what the problem is with inflation. Professor Jones, *you* have the answer.' "[29] Although the press is obligated to express accurately the views attributed to you, it has no obligation to say that you are right.

No matter how skillfully scientists present information to reporters, stories sometimes contain mistakes. "Save your complaints for the errors that count," Robert M. Byers of the MIT news office emphasizes. Unless a piece contains an inaccuracy that could immediately endanger the public, taking some time to cool down is wise. Perhaps show the story to, or discuss the broadcast with, a public information specialist or a colleague experienced in dealing with the press.

If you express an objection, generally do so directly to the reporter. One tactful approach is to preface the criticism with positive sentiments ("I was glad to see your story about my work in yesterday's *Herald*. However, I was a bit concerned that . . ."). Upon learning of objections, journalists may proceed in various ways. If they have misstated important facts, they may arrange to correct them in print. If they have omitted major aspects of topics, they may prepare new stories filling the gaps. In other cases, the feedback merely helps keep reporters from making the same mistakes again.

Occasionally, conveying criticism to the editor instead of, or as well as, the journalist can be in order. For example, you may have been unable to settle a matter satisfactorily with the journalist, the error may have been so serious (or the errors so frequent) that you feel that the editor should know, or you may have learned that the problem resulted from an editorial decision. If the response is still unsatisfactory once you have talked with the editor and if the publication or station has an ombudsman or the equivalent, you may contact that individual as a further recourse.

Especially if you object largely to the emphasis of a story or to the opinions conveyed, writing a letter to the editor (or preparing its broadcast equivalent) can be the most appropriate option. A similar tactic is to compose an Op-Ed piece (i.e., an article to be published opposite the editorial page). Various scientists took such approaches when, in late 1981, *The Washington Post* and other newspapers published an exposé of research on cancer drugs. A letter or Op-Ed piece may reach not only editors and reporters but also the public. However, so many of these items are submitted that a relatively small fraction make it into print or onto the air.

"We don't get any feedback unless we do something wrong," one veteran science writer observes. If you admire a story on your field or work, consider letting the journalist know. A lavish tribute is unnecessary; a sentence or two will do. Such praise generally is most welcome in what has been called "a business of very fragile egos." Giving it encourages journalists to cover science well.

As implied above, objections usually should be expressed orally and only to those immediately concerned, so that they leave the fewest unpleasant traces once the episode is done. With praise, the opposite is true. If a story is exceptionally fine, write the reporter a note, which can be savored, reread, and shared; send a copy to the editor. Perhaps submit a brief letter for publication as well. Let colleagues know that you respect the piece; they then may be more willing to work with its author, and with the media in general. Also recommend the story to non-scientists, for you can thus help to inform the public about science.

Responding to Interest Generated by a Story

Upon encountering a story in the mass media, various parties may seek to learn more. Journalists, members of the public, and even scientists who were previously unaware of your work may contact you once you receive attention in the press. Often, a public information office can aid in dealing with what, if an item is especially newsworthy, can be a veritable barrage of calls and mail.

Much of this book addresses working with journalists, and you already know how to provide information to peers. But how can you respond most easily and effectively to requests from the public? A few hints may help:

- Anticipate the most common questions and prepare answers to them.
- Be ready to recommend popular and semi-technical readings that discuss the topic more fully than the media coverage did and that address related subjects.
- Perhaps prepare a form letter that can answer many of the requests. A public information office may be able to aid you in doing so.
- If you are getting many phone calls, consider setting aside a specific part of the day to answer them. Maybe even install a recording to answer calls.[5]
- If responding to the publicity is interfering greatly with your work and that of your staff, try to get extra personnel to help with preparing replies.

Such measures can facilitate presenting science to a highly attentive audience.

"These people do go away. Then you can get back to what you do," one reporter notes. Attention in the news is fleeting, but its impact is not. The public knows more, and perhaps cares more, about science than before. Journalists know more about the topic and may have new contacts in the scientific community. The scientists involved may understand more about how the press works. In journalism as in science, learning is a continual process for all involved, and opportunities lead to other ones. Such are the roots of progress in presenting science through the mass media.

PART III

PRESENTING SCIENCE DIRECTLY TO THE PUBLIC

Why should scientists present science directly to the public? Why not leave such activity to professional science communicators? As scientists, we are the ones best acquainted with both the process and the products of science. In addition to our expertise, we have the authority to speak out. Because we do not depend on popular science communication for our livelihood, we sometimes can prepare our pieces more slowly and carefully than a deadline-ridden reporter or a hungry freelancer can. Finally, some of us find that we enjoy presenting science to the public; this elusive element is often what transforms such activity from something that we merely acknowledge as worthwhile to something that we actually do.

Chapter 11

Breaking In:
Articles, Books, and Broadcasting

In science communication as in science, opportunities most often fall to those who know the proper outlets for their work, prepare strong proposals, and strike sound agreements. This chapter therefore addresses the science writing equivalent of grantsmanship. First, as a general model, it discusses arranging to write an article; it then briefly comments on book publishing and broadcasting.

Breaking In: Articles

For those trying to publish popular articles, a basic question is "Where?" A logical way to begin seeking an answer is by recalling which publications have dealt well with the relevant fields. Browsing in a library may also help. Another resource is *Writer's Market*,[79] an annual volume that lists, by subject area, various (although not all) magazines accepting freelance work. Among the science-related categories are health; home computing; nature, conservation, and ecology; science; and social science. For many of the listings, the book provides such information as size of circulation and kind of audience, types and lengths of pieces sought, and payment rate, in addition to the publication's address and the editor's name.

Of course, such sources as memory and *Writer's Market* are not enough to consult before proposing an article. Cary Lu, executive editor of *High Technology*, emphasizes that prospective authors "should read the magazine very carefully and try to understand who the magazine is addressing"; other editors stress similar points. Among items to check when inspecting a publication are its apparent function; the subjects, types, and lengths of articles; the style, tone, and difficulty of the writing; and the seeming audience. Also ascertain that you have not chosen such an appropriate publication that it has already published essentially the same story as you would propose.

As implied above, publications vary in their willingness to accept freelance articles. At one extreme, the science sections of the major news magazines are written entirely by their own staff. At the other, some university magazines vigorously recruit pieces by faculty. The blurbs that many magazine and newspaper articles run regarding their authors are one clue as to whether they might be interested in your work.

Publications also differ widely in how much they pay for articles. Table 4 presents examples of payment rates at various potential outlets for science stories.

After identifying a publication that seems appropriate, contact the editor. "We don't like to get finished articles unsolicited," says Susan Williams, an assistant editor of *Science 83*. "We just don't have time to go through a whole manuscript." Consulting the editor before writing a piece is most efficient for an author as well.

Initially, writing even as much as a proposal can be inefficient. Why spend the time preparing one when, unbeknownst to you, the editor may already have commissioned a story on your topic, may have a policy against accepting anything on it, or currently may not be looking for any freelance pieces? A brief preliminary letter can thus be the best approach. In the letter, summarize your background in science and any experience in popular writing; perhaps include a brief résumé or curriculum vitae. If you are interested only in a particular story, mention your idea. Otherwise, describe the general types of articles that you hope to write, and propose a few topics. If you have written for the public, enclose one or a few samples of your work.

The editor can quickly peruse such material and, if appropriate, pass it on to another member of the staff. If the publication is uninterested, you have lost little effort. If, however, the answer is encouraging, you most likely will receive clues on how to proceed.

Especially if you are inexperienced in writing for the public, the editor may ask for a proposal—or, in journalists' terms, a *query letter.* The basic elements of such a letter resemble those of other proposals that scientists write. A query letter may, for example:

- briefly state the gist of the article
- say why the idea is worthwhile (e.g., what makes it newsworthy)
- note why the author is qualified for the task
- describe the proposed article in more detail (e.g., outline it)
- discuss plans for preparing the piece (e.g., timetable, length of piece, and major sources to be used)

Samples of an author's work, if not sent beforehand, may accompany a

TABLE 4. *Examples of outlets for freelance articles*

Publication	Length of article	Payment†
Publications specializing in science		
American Health	150–500 words	50¢ per word and up
	750–3,000 words	50¢ per word and up
Byte Magazine	variable	$50 per printed page
High Technology	300–800 words	$250 to $800
	3,000–4,000 words	$1,800 average
*Natural History**	2,000–4,000 words	$300 to $750, plus extra for photos used
*Psychology Today**	3,000 words	$550
Science Digest	600 words or less	about $130
	800 to 1,200 words	$500 to $800
	2,000 words or more	variable
Science 83	up to 1,200 words	varies according to type of article; often about 60¢ per word
	3,000–4,000 words	usually $2,000
Scientific American	4,000–5,000 words	$1,000
Sci-Tech (weekly supplement to *The Boston Globe*)	about 1,000–1,500 words	$100 to $500
	up to about 2,500 words	variable
Technology Illustrated	1,500–2,000 words; short pieces also	$50–$2,500
Technology Review	2,000 words	$100 to $200
	5,000 words	$250 to $500
3-2-1 Contact* (children's magazine)	700 to 1,000 words	$150 to $300
*Publications not specializing in science**		
The Atlantic Monthly	1,000–6,000 words	$200 and up per printed page
Boston Globe Magazine	2,500–6,000 words	$500 to $750
The Christian Science Monitor	usually under 800 words	$50 to $75
Family Circle	1,000–2,500 words	$250 to $2,500
National Geographic Magazine	2,000–8,000 words	$3,000 to $8,000 and up
The New York Times (Op-Ed Page)	about 750 words	about $150
The New York Times Magazine	about 4,000 words	$850
Parade	800–1,500 words	$1,000 and up
Playboy	4,000–6,000 words	$3,000 and up
USAIR Magazine	1,000–3,000 words	$400 to $1,000

* Data on these publications were obtained from *1983 Writer's Market*.[79]
† Amounts paid sometimes differ from those publicly stated.

query letter. "We don't like to be inundated with clips, but we rarely buy things from people who are unpublished," Williams notes. "We like some evidence that they have written before, though it's not imperative."

Of course, gear the proposal to the target publication. Make it meaty but brief—in general, two pages at most.[54] Perhaps needless to say, be sure that the proposal is well written, for here is a chance for the editor to see how you write. In particular, take care to keep the language non-technical, for editors often worry that a scientist will be unable to communicate to a general audience.

The more stories an author writes, the simpler the query procedure tends to become. An editor pleased with your past work may want only a brief note proposing a story; even a phone call might do. Editors of other publications, too, may seek briefer proposals once you establish a good-sized portfolio, a reputation in science writing, or both.

In fact, as your science writing becomes known, editors may approach you. Such flattery can be hard to resist, but many of us have lived to regret accepting an assignment uncritically. Among basic but oft-neglected questions are the following: Are you comfortable with the publication? Does its staff seem pleasant to work with, and what is the scuttlebut about how its editors treat authors and their manuscripts? Does the suggested topic suit your background and interests? (If not, consider suggesting other subjects; you are in a fine position to propose ideas if an editor has sought you out.) Do you really have time for the assignment? What impact would the prospective article have? And, yes, how much will you be paid?

Once an editor has accepted a story proposal or has approached you, it is time to agree on the details. Like a query letter, the agreement is likely to be formal if an author and an editor have not worked together much before; it may be a detailed letter from the editor or an actual contract. Later, when rapport and trust are well established, the agreement may take briefer and more casual forms, becoming a simple note or merely being conveyed by word of mouth.

Whether the agreement is formal or informal, it should specify matters such as the following:

- *topic, slant,* and *format* of article
- *length* of article
- *deadline*
- *amount of payment*
- *timing of payment* (e.g., upon receipt of the manuscript or only once the article appears in print)
- availability of *reimbursement for expenses* (e.g., for travel and long-distance telephone calls)
- availability of an *advance* to cover such expenses

- availability of a *kill fee* (i.e., partial payment if a commissioned article is received but not run)
- *right to review editing* (i.e., whether the author will have a chance to comment on editorial changes; also, whether the author's remarks will be viewed only as suggestions, as is usually so, or will represent the final say)
- holder of *copyright* on the piece

Clarifying such items at the start aids in preparing the article and helps prevent misunderstandings later on.

Breaking In: Books

"In years past, editors and publishers say, most science books for the public were written by laymen with the ability to make abstruse subjects understandable, whereas today they are increasingly written by scientists themselves," a recent article in *The New York Times* notes.[57] Book publishing is another option for scientists wishing to present science directly to the public. In fact, a book may be easier for a scientist to get published than an article is,[56] and thus it may be a good way to break into print. Of course, a book is much more work than an article to prepare.

Books arise from many sources. The idea for a work can spring nearly full-blown from an author's brow. Series of popular lectures by scientists sometimes give rise to books. So do series of articles originally published in periodicals (e.g., the essays of Lewis Thomas and Stephen Jay Gould). And when a scientist and a professional science writer hit if off especially well, they sometimes collaborate on a book.

The principles for finding an appropriate book publisher are much the same as for locating an outlet for an article: note who has published works similar to that which you envision, check where opportunities might exist, and then submit a proposal. Writing brief preliminary letters of inquiry is especially wise in the case of books, for book proposals are much longer and more time-consuming than query letters for articles; a typical book proposal contains such elements as a detailed outline, a sample chapter, a curriculum vitae or biography of the author, and a discussion of such items as the potential market for the work. Whereas authors can validly send short initial inquiries to several publishers simultaneously, they probably should not submit full proposals to more than one at a time without stating that they are doing so.

Although normally a book contract is much more extensive and detailed than the agreement regarding an article, it centers on the same major types of elements: what you must provide by when, and what you will get in return. William Bennett, a physician who has experienced book publishing from both the editor's and the author's ends, notes that a common scale of book royalties is 10 percent of the list price (i.e., the price on the book jacket) for the first 5,000 copies sold, 12.5 percent for the next 5,000, and 15 percent for the rest. "It's commonly said that if you sell 5,000 copies, you've done pretty well," he adds.

Publishing companies and general readers are not the only ones calling for non-technical books by scientists. The Commonwealth Fund has established a program mainly to sponsor scientists in writing books "to inform the lay public of the discoveries and work now under way on the frontiers of science." Information about this opportunity is available from The Commonwealth Fund Book Program, Memorial Sloan-Kettering Cancer Center, Room 604 Schwartz Hall, 1275 York Avenue, New York, New York 10021.

Breaking In: Broadcasting

How can a scientist enter broadcasting on the side? "You get the breaks. It's show business to a degree," says biologist Gene Kritsky, whose minute-long science segments appear on WOWO Radio in Fort Wayne, Indiana. For Kritsky, who had written newspaper articles and press releases while a graduate student, the unexpected break came when, after interviewing him on a farm program, the station invited him to do some broadcasting himself. Others tell analogous stories: one physician who appears on a television magazine program had happened to treat members of its staff, and another was recommended by a friend.

Breaking into broadcasting is, however, much more than a matter of chance. Broadcasters offer invitations, and keep them open, not at random but rather to scientists who can present science effectively to the public. Also, one need not wait for chance to call. Opportunities in broadcasting often go unrecognized by the scientific community, Carol L. Rogers of the American Association for the Advancement of Science emphasizes. Rogers notes that many local stations may be pleased to give scientists with "free time and free information" some access to the air. She also mentions that scientists can work as consultants to science-related television programs.

For a busy scientist, turning out polished broadcast segments on a rigid schedule can be impractical. However, more flexible and less demanding alternatives exist. One such option is for a scientist to join a talk show's stable of frequently appearing guests. Another is for the scientist to host an interview program or interview segment, especially if staff members at the station do much of the legwork for it.

Breaking into broadcasting can mean breaking out of a scientist's usual role. "A key to it is not to turn these things into lectures. Then they get boring," says physician Leonard A. Katz, a frequent guest on WKBW television's talk show *AM Buffalo*. "Be able to take criticism from your audience and from the professionals at the station. Don't take it as an affront to your education," Kritsky adds. "If someone has 25 years of experience in broadcasting, there's something to be learned from that individual."

Dealing with the Business Side

Scientists may not view their popular science writing as a business. Their principal employers, however, may. Their editors will. The Internal Revenue Service will, too. Awareness of a few basic items will aid in dealing with the business side.

Some organizations, such as companies and government agencies, prohibit or limit outside work. In particular, some have clearance procedures to help prevent conflicts of interest. Finding out the policy before looking into science-writing opportunities can prevent problems later on.

Dealings with editors should follow the same principles as other types of business. Of course, note deadlines and observe them; if difficulties arise, inform the editor right away, so that adjustments can be made.

As for the IRS, record any financial transactions regarding your work. Make a note of any income from freelance writing as soon as it arrives, and report it on Schedule C of your tax return. Also, promptly record any business expenses that you plan to deduct. To simplify completing your tax return, list deductible expenses in the same categories as on Schedule C (for instance, dues and publications, office supplies and postage, travel and entertainment, and utilities and telephone); be sure to keep receipts.

With sound grantsmanship, scientists can often win opportunities to present science to the public. So, now to discuss the work itself . . .

Chapter 12

Interviewing and Other Basics

Samuel Johnson observed that "The greatest part of a writer's time is spent in reading, in order to write; a man will turn over half a library to make one book." Likewise, in presenting science to the public, composing the actual product is only a fraction of the work. This chapter therefore addresses two key aspects of science communication: choosing topics and gathering information. It devotes particular attention to conducting and using interviews.

Choosing Topics

Scientists can identify topics for stories in many ways. Like professional reporters, they can draw ideas from major scientific journals and conferences, from science items in the mass media, and from events in the general news. In addition, they have greater access than journalists to many sources of story ideas. These sources include one's own work and that of one's colleagues, meetings and publications too specialized for even full-time science writers, and various contacts in the scientific community. In assessing which topics are appropriate for lay audiences, reviewing the elements of newsworthiness (see pages 30–32) can be worthwhile.

"Write about what you know about" is a basic maxim. Knowing too much about a topic may, however, interfere with presenting it to a general audience. Scientists sometimes write their best popular stories about work well removed from their own—for example, when they are approaching an unfamiliar area that intrigues them, or when they come to a writing project with a question to explore rather than an answer to convey. By allowing both emotional and intellectual perspective, such distance aids in preparing work that outsiders find relevant and clear. It also helps in interviews; in the search for broad insights and crisp quotations, one can pose basic questions more easily to distant colleagues than to one's closest peers. Finally, when a project entails discovery, not only is the process most exciting for the author; the product is likely to be most exciting for the audience.

Gathering Information

Good science communication, like good science, usually entails gathering much more information than will appear in print. As one science writer puts it, "Good writing can be measured by the good stuff it leaves out."

To the creative, many sources of information are available. Scientific publications, your colleagues, and your own experience may yield plenty of scientific information to use in a popular piece. Additional resources, however, can enrich and enliven a work. For instance, if a phenomenon (be it infertility, automatic teller machines, or the new show at the planetarium) affects community members, consider interviewing some who have experienced it. Think about tracking down some historical background. Maybe see how literature, movies, and song have portrayed your topic. If a subject has policy aspects, consult relevant government agencies, congressional offices, and interest groups. Perhaps note how popular writings and broadcasts have already presented your topic (beware of doing so initially, though, for fear of influencing your own approach too much); to identify these items, use such aids as the card catalogue at a large public library, the *Reader's Guide to Periodical Literature*, and *The New York Times Index*. Use your imagination— and consult your reference librarian.

Deciding Whether to Interview

As Chapter 5 discusses, journalists do interviews for various reasons. When you, as a scientist, are preparing a story, some of these reasons may not hold, especially if you are writing about your own field. You can easily understand the science; no one need translate it for you. You can tell whether a story idea is scientifically sound, and you can place a topic in perspective, both scientific and general. You may be a recognized authority yourself; stating your own opinions may suffice.

Nevertheless, interviewing others may well enhance your story. You may, for example, be venturing beyond your own speciality, seeking information not yet in the literature, or looking for other scientists' opinions to cite. You may wish to enliven an article with quotations, including the sorts of statements that seem witty in quotation marks but trite in an author's own prose. You might want to include views of non-scientists, such as government officials and community members. Or you may simply be writing about another scientist. Interviews serve many important functions, and editors complain that failure to conduct them is among the most common problems when scientists write popular articles on science.

Facing Issues of the Scientist as Interviewer

Scientists' failure to conduct more, and more successful, interviews may stem in part from special issues that face the scientist as interviewer. One challenge is to obtain scientific information in lay terms from peers. If you are interviewing colleagues who know you as a scientist, they may answer your questions in jargon (and then perhaps embark on tangents that interest only those in your field). One approach to this problem is to ask such questions as: "How, in common language, would you describe . . . ?" and "If you were explaining this to a high school class, what would you say?"

When you are interviewing a scientist who does not know you, one dilemma is how much of your background to reveal. On the one hand, not mentioning your scientific training may win you simpler, more quotable statements. It may also save you from questions about why you are wasting your time on literary pursuits. On the other hand, indicating that you are a scientist can be crucial in gaining cooperation, especially if a source distrusts reporters. Also, consciously deceiving a source not only can be embarrassing should your true identity later emerge but is downright unethical.

How much to say about yourself depends in part on an interview's purpose and length. If you are only calling to ask a few straightforward questions, your training may be irrelevant; stating your name and identifying your project may suffice. However, if you are asked about your qualifications, if a source seems reluctant to deal with "mere" reporters, or if you will be conducting an extensive interview, the best approach may be to mention your training but play it down. For example, you might say, "I have a background in such-and-such field, but so-and-so subspecialty isn't my own area. I hope that you won't mind if I start with some pretty basic questions just to make sure that I have things straight and can present them accurately to a general audience." Playing a little dumb can be smart in an interview.

Another question is how to identify yourself when interviewing non-scientists. Mentioning your background can inspire confidence, but it also can intimidate interviewees. What to say, and how, is a matter of judgment here, as above.

Special concerns arise if you are writing about medicine and wish to interview individuals with a particular condition or disease. Approaching your own patients for interviews is not only awkward but potentially unethical. And it is certainly unethical for another clinician to identify patients without their consent. One alternative is for a colleague to mention your needs to patients in the appropriate categories; those interested in talking with you can then call you directly or grant permission to release their names. Another option is for you to contact a relevant organization (for instance, the Arthritis Foundation or the American Cancer Society) and ask for patients to interview; such volunteers are likely to be a skewed sample, but at least they are probably eager to share their experiences with your audience. Also, consider interviewing friends, acquaintances, colleagues, and family.

The scientist who assumes the role of interviewer should remain alert to various potential traps. Take care to pursue the information that the public wants and needs, rather than concentrating on what interests you as a scientist; consulting your editor and representative audience members beforehand can aid in deciding what to ask. Avoid the temptation to display your own knowledge rather than obtaining that of the source. Finally, beware of lapsing into scientific jargon when talking with sources: clear common language not only is essential in interviewing non-scientists; it also serves as a model when you are seeking non-technical statements from scientists.

Conducting the Interview

"The degree to which you prepare for it will determine how successful the interview is," writes Paul Desruisseaux of the University of Southern California in "The Ps and Qs of Q and A."[22] Digesting background materials, defining the purpose of the interview, and deciding what sorts of questions to ask are all seemingly obvious but sometimes neglected parts of preparing well. Letting your sources know your needs in advance, so that they too can prepare, can also aid in interviewing.

The choice of medium is another element of success. In general, a face-to-face meeting yields the best rapport and the most material. However, the telephone may be the only reasonable option when time is short, your questions are few, or your source is far away. Sometimes even otherwise a telephone interview is preferable. Your invisibility can improve the interview: some subjects speak more freely when they cannot see your reactions and cannot tell when you are taking notes; also, you may conduct interviews most effectively when spread out comfortably at your desk. A third alternative is to obtain information by letter; if a request is narrowly focused and a source is hard to reach, correspondence may be the best approach.

Success also depends on recording the information in a convenient, reliable way. If you are a confident and adept note taker, pen and paper may suffice; in fact, some writers prefer written notes, which are easier than tapes to consult. In general, write quotable quotes verbatim but summarize other material. If your pen falls behind but you fear that asking the speaker to slow down will disrupt the flow of the interview, consider alternating major questions with minor ones, so that you have time to write. For convenience later, underline or star important items.

As John Brady discusses in *The Craft of Interviewing*,[13] tape recording an interview can have various advantages. For example, taping aids in capturing verbatim quotations, allows you to ask more questions per unit time than note-taking does, and lets you concentrate on the interview. Also, carrying a tape recorder can be much easier than writing if you are touring a laboratory, walking in the field, or otherwise on the go. Beware, however, of relying on tapes alone; if possible, write down the main points made and their approximate location on the tape, any important quotations, and perhaps topics about which to ask more. Such note-taking not only helps you to conduct the interview smoothly and to use the tape efficiently afterward; it also provides back-up in case the recording fails.

Of course, effective interviewing technique is also basic to success. In talking with a source, the following suggestions may be of help.

- Put the source at ease, for instance by beginning with small talk. "If you don't have rapport," science writer Mitchel Resnick observes, "the answers won't have any spark."
- Ask one question at a time to make sure that each is answered.
- Work from general to specific questions to prevent closing off valuable avenues. Also, beware of "leading questions" (i.e., questions that "put words into the speaker's mouth"); they usually yield little information (and even less in the way of quotable quotes), and they may antagonize a source.
- To prompt a source when an answer seems incomplete, perhaps pause for a moment or echo the last few words.
- Stay alert for unexpected items worth pursuing.
- Consider reiterating the main points that the subject made. Doing so not only increases accuracy but also may inspire the source to elaborate.

- End, as well as begin, with general questions. For example, ask the subject to sum up the most important point or to state anything else especially important for you to know. Useful perspectives and quotable quotes are likely to result.
- Put the subject at ease again when the interview ends, and perhaps offer to send a copy of the published article.

While the interview is fresh in your mind, review your notes or tape recording, and put the information in usable form. If new questions occur to you, consider conducting more interviews with either the same source or other ones. In journalistic as in scientific research, reflecting on answers often leads to further inquiry.

Using the Interview

An interview, like many another expedition, yields some debris to discard, some resources to process, and some gems to polish and then use. Include only what will enhance the story. If the entire interview may be of historical interest, consider offering your notes or tapes to the appropriate archives.

No matter how skillful the interview, some information will be irrelevant. Other material will aid you as background but does not belong in the story itself. Beware of the temptation to include every hard-won bit.

Not all material worth including is worth quoting. The general rule is, "If you can say something at least as well yourself, do so." If neither the content nor the form of a statement is novel, the best approach is usually to present the content yourself, without mentioning the source. You would not formally attribute the statement "Jupiter is the fifth planet from the sun" or "The human heart has four chambers" to a conventional encyclopedia; you need not attribute it to a walking one.

If the content or general form of information presented by a source deserves acknowledgment but the original words are too technical or otherwise unquotable, consider using what is called an indirect quotation. Say, for instance, "Professor Lewin explains that the new process consists of four basic steps. The first, she notes, begins . . ." Similarly, use an indirect quotation to convey a speaker's opinions if the original words do not work; you can paraphrase the entire statement, or you can paraphrase sections and quote the quotable parts.

Journalists poorly versed in science sometimes toss a jargon-filled statement into quotation marks because they sense that it is important but they cannot understand it well enough to translate it. As a scientist, you are unlikely to fall into this trap. However, you may be so familiar with the jargon that you are unaware of it. Thus, check statements carefully for jargon. When you find it, paraphrase rather than quote.

When, then, should you quote your sources verbatim? Mainly when their speech will impart life, or force, or both. Use a quotation when the source's words are vivid and apt. Quote experts to add authority. When people describe their experiences, add authenticity by using their own terms. Sometimes, though, even if quotations do not meet such criteria, inserting a few of them can enhance the texture of a piece and add human warmth.

If you use quotations, the question may arise of how much to edit them. William Zinsser, the author of *On Writing Well*, answers with a short phrase: "brevity and fair play."[92] Spoken language is flabbier than written; it is more repetitious, and it contains constructions that prove incorrect, or at least seem odd, in print. You will serve both your readers' convenience and your sources' images if you delete the um's, er's, and redundancies and correct grammatical errors. Beware, however, of any changes, including omissions, that may distort what was said.

Using quotations to full advantage entails skillful breaking and attributing. If you are interrupting a quotation, do so at a logical pause. Say: "Food," Bennett and Gurin write, "is a red herring."[10] Not: "Food is a red," Bennett and Gurin write, "herring." Editors differ among themselves on the need to vary wording by using synonyms for "say." Words like "argue," "explain," and "complain" can enliven a piece; an occasional "state" or "note" or "add" can lend welcome variety. Do not feel, however, that you must set off each quotation with a different verb, so that you have your poor source "expound," "declare," "declaim," and "ejaculate."

The topic is chosen, the information is gathered—and a publication or station is waiting for the story. Now comes the actual writing, the topic of the next three chapters.

Chapter 13

Writing the Piece:
Beginning and Beginnings

Roger Swain, biologist turned science writer, notes that he spends five days composing a typical 2,300-word article. By the end of the first and most agonizing day, he has come up with his first sentence. By the close of the second day, he has completed the first paragraph. The third day, he finishes the first page. In the remaining two days, he writes the rest of the piece.

Beginning and beginnings are usually the hardest part of science writing. Yet at some time during their careers, most scientists face starting to prepare a popular or semi-technical piece on science. Of scientists surveyed at two universities, nearly one third said that they "had written at least one story themselves (perhaps a press release) for mass media dissemination."[28] Much more often than they write popular articles, scientists produce textbook chapters for non-specialists, lectures for students and the public, handouts for classes, instruction sheets for patients, reports for administrators, and other materials for non-peers. In composing all such products, many of the same basic principles hold.

This chapter and the next two deal with various of those principles, especially as they pertain to presenting science to the public. Good science writing, however, must first be good writing *per se*. Thus, scientists interested in expressing themselves effectively should consult such general works as *The Elements of Style* by William Strunk, Jr., and E. B. White[82] and *On Writing Well: An Informal Guide to Writing Nonfiction* by William Zinsser.[92]

Choosing the Proper Setting

I must confess that when I started preparing this book, I decided to improve my writing habits. So, when I began the actual writing, I gathered my notes and set off for the library. I arrived there early in the morning. And there I sat. Sat, not wrote. After another such frustrating episode, I reverted to my old mode: working at the kitchen table. Then the words began to flow (from my favorite mechanical pencil, onto the narrow-lined paper that I prefer), generally at several pages per stint.

As Zinsser states:

> ... There are all kinds of writers and all kinds of methods, and any method that helps somebody to say what he wants to say is the right method for him.
>
> Some people write by day, others by night. Some people need silence, others turn on the radio. Some write by hand, some by typewriter [or word processor, one might add today], some by talking into a tape recorder. Some people write their first draft in one long burst and then revise; others can't write the second paragraph until they have fiddled endlessly with the first.[92]

As a scientist, you probably have been writing throughout your education and your career. By now, you most likely know under what conditions you write best. If not, keep experimenting. And, as Zinsser advises, write under whatever circumstances are effective for you.

Getting—and Keeping—Going

Beginning a writing task is much more difficult than starting to wash the dishes or mow the lawn. One major reason is that it entails not only overcoming inertia but also planning the project. In formulating ideas for a piece of writing, various methods can be of help.

One approach is to become thoroughly familiar with the material and the task—and then to turn your attention elsewhere. In science, intuitions often arise during apparently idle periods following intense work[11]; as J. Robert Oppenheimer put it, "Theoretical insights flourish best when the thinker is apparently wasting time."[34] Similarly, escaping a writing project for a while can aid in obtaining thoughts for it.

Another strategy is to "talk it out." When Paul de Kruif, a bacteriologist who became a best-selling author, was beginning his writing career, an editor advised him to tell train conductors about his topics. "When you've hooked them," the editor said, "write it like that."[9] You need not track down a train conductor; but, as Rudolf Flesch notes in *The Art of Readable Writing*, "The most widely used device for getting ideas in shape is to buttonhole some unsuspecting victim—the kind of person who is apt to read later what you have written—and to rehearse your ideas aloud."[34] A related tack is to envision a member of the audience and then dictate into a tape recorder.

Making one or more lists is a third method. Jot down points that you wish to include. Then a way to begin and structure the piece may suggest itself. Either immediately or after making a more formal outline (a step to consider no matter how you crystallize your ideas), you can proceed to write.

Finally, a fourth tactic is merely to overcome your inertia and start to write. Once something is down on paper (or up on the terminal), an approach may become clear. Thus, even if not used in your final draft, your initial words may serve an important role.

Once you get going and have a direction (and perhaps a map as well), the writing usually becomes easier. If you are preparing a short item, you may be able to complete a draft in a single stretch. Otherwise, where should you pause? The "logical" sort of stopping place (for instance, between major sections of an article, or where you are not sure what to say next) may not always be so logical after all. Sometimes, it is better to break off in the middle of a passage while you are still going strong. Then, in resuming your writing, you can harness the energy already amassed.[69]

Beginnings (and Middles and Ends)

In part for the same reason that starting can be the hardest part of writing, the opening paragraphs are often the most difficult to compose. "If you haven't written a good lead, you don't understand what the article is about," Roger Swain remarks. Other factors, too, make beginnings a challenge to produce. Especially if audience members lack preexisting interest in a topic, a story's opening must entice them to read on. Thus, notes William Bennett, who edits *The Harvard Medical School Health Letter*, a good lead should not only indicate the topic and thesis of an article but also "create some kind of question or tension in the reader's mind." Furthermore, a lead should show that a writer's style is interesting and readable.

Both in print and orally, various strategies can help you to snag an audience. One tactic is to build on the public's current interests by using a "news peg." For example, a *New York Times* article begins:

> The paralyzing nature of the snowstorm that hit much of the East Coast in recent days caught New York and New England by surprise, showing once again that a small meteorological error can make an enormous practical difference.
>
> Weather forecasters have become more accurate in making short-term predictions, and more sophisticated equipment and techniques may help to make them even more accurate in the future. But the complex factors that go into creating weather patterns indicate that true accuracy is still far away.[12]

Even when topics are essentially unappealing, you can draw interest by immediately noting their importance to the public, as Jane E. Brody does in one of her "Personal Health" columns:

> In developing countries, where sanitation and personal hygiene often leave something to be desired, diarrhea is the leading cause of childhood death and stunted development and a frequent hazard to foreign travelers. Even in the United States and other advanced countries diarrhea is a significant cause of illness, among Americans second only to respiratory infections.[14]

Impressive (or otherwise interesting) facts, as well as intriguing questions, can also help attract an audience. For instance, an article in *Smithsonian Institution Research Reports* starts:

> Imagine reading a novel in Los Angeles from as far away as Boston. This is the level of resolution—one ten-thousandth of an arcsecond—possible with a radio astronomy technique called Very Long Baseline Interferometry, VLBI.[68]

And a brief story in *Science 83* sports the following lead:

> How does an outfielder know where a fly ball is going to come down? According to physicist Peter Brancazio, he plays it by ear.
> As a player tracks the ball with his eyes, he also moves his head. This head movement, says Brancazio, is picked up by organs in the inner ear—the same organs that provide the brain with the information needed to maintain balance. By monitoring the changing tilt of the head, the brain is able to calculate the path of the ball.[46]

An especially effective way to start can be with human interest, by telling a story about a person. The individual may, for example, be a scientist (perhaps yourself) at work or a community member who uses the technology or suffers the disease to be discussed. A clever twist on this approach is the following lead that Andrew Pollack wrote during the summer of 1982:

> Pretend that Edward Thames, a college student, is traveling around California on vacation when an emergency arises at his home in New York. His mother dials a local phone number and transmits a message. Almost instantaneously, Edward pulls out his pocket pager and reads the message. "E.T., phone home," it might say.
> Such a scene is not likely to occur today, but could occur in several years. Radio pagers, popularly known as beepers, are undergoing technological developments that promise to vastly expand

their market beyond hospital employees and equipment repair-men.[67]

Granted, narrative leads may be getting hackneyed, and sometimes the tale is too tangential to the subject matter. Used sparingly and skillfully, however, such beginnings can be an excellent means of attracting an audience.

Once underway, a piece should, of course, proceed along orderly, easy-to-follow lines. For example, it may provide a chronological account or an amplified list, compare and contrast alternatives, or describe cause and then effect. Lengthy articles often combine several such strategies.

However an item is structured, it should present information at a *pace* that is comfortable for the audience. When offered too close together, unfamiliar ideas can overwhelm recipients. Therefore, intersperse other material (such as examples, restatements, quotations, explanations, and anecdotes) according to readers' or listeners' interests and needs. Trying out an item of science writing on representatives of the intended audience will aid in deciding whether the pace is appropriate.

Other than the beginning, the hardest part to write is often the end. But how you started may suggest how to stop. If the audience was interested from the outset and wanted you to plunge right in, it may want you to jump out without much ado once done. If, however, you have been attracting your spectators' attention with goodies right along, serve something tasty for dessert. For example, finish with a surprise, a lively tale, a zippy quotation, a clever turn of phrase, an impressive fact, or a bit of irony. In particular, consider unifying the piece by circling back to the idea with which you began. Speaking of which, Roger Swain likes to end his articles with outrageous puns.

Chapter 14

Communicating Crisply

Throughout our educations and careers, many of us have associated verbosity with success. Not a few of us recall teachers who apparently would count the pages in a term paper, add to that the number of words at least five syllables long, and then multiply by a constant to calculate the grade. Many of our peers still think that lengthy terms and convoluted sentences mark the scholar or professional.

Long words and sentences, however, make writing hard for anyone, including colleagues, to read. Couple them to content that is unfamiliar and technical, and they are likely to drive the public away. Thus, an important principle of writing for general audiences is to keep most of your words and sentences short. This chapter presents advice on doing so. Some of the suggestions apply to communicating crisply not only in print but also orally.

Following the Basic Principles

Awareness of several basic principles can aid in communicating crisply. The first guideline is to *discard needless words*. Much writing, especially in science, contains useless verbiage. Trimming this fat away (see Table 5 for examples) makes prose quicker, easier, and more pleasant to read.

In addition, *replace long words with short ones* when possible (see Table 3, page 12). One helpful tactic is to imagine speaking to the intended audience in appropriate language. A related aid is to read the writing aloud; if any words stick in your throat or seem unsuited for the lay ear, search for substitutes. Also, *condense wordy phrases*, as shown in Table 6.

TABLE 5. *Examples of removing needless words*

Original version—contains needless words (underlined)	Crisper version—needless words have been removed
opportunities in the field of engineering	opportunities in engineering
The device is of an efficient nature.	The device is efficient.
green in color	green
count the number of fruitflies	count the fruitflies
in order to help	to help
anyone who is interested in such problems	anyone interested in such problems
a protein that is called the LDL receptor	a protein called the LDL receptor
Such injury is a very rare event.	Such injury is very rare.
whether the services should or should not be available	whether the services should be available
to determine whether or not the invention works	to determine whether the invention works
The fact of the matter is that no safe test exists.	No safe test exists.
because of the fact that this liquid is denser than water	because this liquid is denser than water

Fourth, *use verbs rather than nouns and adjectives made from them.* Scientists often favor nouns and adjectives rather than the verbs to which they correspond. For instance, they say "experienced an increase" rather than "increased," "the effects of the introduction of the technology" rather than "the effects of introducing the technology," and "is hopeful that" rather than "hopes that." The resulting prose often is wordy and hard to read; because verbs give language its action, such writing sounds static as well. For clues useful in identifying "nounification" and examples of how to correct it, see Table 7.

To make your writing lively and brief, also *use active voice* in most sentences. For example, instead of saying "The housing problems of the elderly are addressed by the study," say "The study addresses the housing problems of the elderly." Passive voice can, however, be validly used when the object of the action should be emphasized.

Furthermore, *say what things are, not what they are not.* Many scientists are fond of phrases such as "not insignificant" and "not inconsistent with." Stating ideas directly makes prose clearer and more vigorous. Thus, especially when communicating with the public, beware of speaking in negatives. For instance, do not say "has not received much attention" if you mean "has received little attention," and do not say "failed to accept the theory" if you mean "rejected the theory."

Beware of sentences beginning "It is" or "There is." Often, you can

TABLE 6. *Examples of condensing wordy phrases*

Wordy phrase	Briefer substitute(s)
a great deal of	much
a variety of	various
as a result	thus; therefore
at high speed	fast
at present	now
at some future time	later
at some point in time	sometime
be detrimental to	harm
does not have	lacks
each of the children	each child
for this reason	thus; therefore
from a medical point of view	medically
generate hard copy	write
have negative effects on	harm
has the capacity to	can
has the potential to	can; may
in most cases	usually
in the course of	during
in the event that	if
in the most effective manner	most effectively
in this manner	thus
is able to	can
is capable of	can
is similar to	resembles
is sufficient to	can
of low cost and high efficiency	cheap and efficient
on an industrial basis	industrially
one of the laws of physics	one law of physics
the majority of	most
this sort of	such
this type of	such
those who have the disease	those with the disease
was responsible for	caused

make such sentences more active and concise. For example, change "It is not necessary to remove the vapor" to "The vapor need not be removed" or (if appropriate) "You need not remove the vapor." Similarly, condense "There is another method that is gaining acceptance" to "Another method is gaining acceptance."

Finally, *consider using clauses beginning with "question words."* Such clauses can add vigor and brevity. For instance, you can replace "to determine the extent of the invention's usefulness" with "to determine how useful the invention is," and you can convert "after describing the process of making a cartoon" to "after describing how a cartoon is made."

DUFFY   by BRUCE HAMMOND

Estimating Readability

By following the above advice, you will not impress those for whom bigger is better in anything, including manuscripts. Fortunately, some formulas work on the reverse principle: the lower a writing sample rates on some composite of such factors as word length and sentence length, the better its "readability score."

One of the most widely and easily used of these readability scores is the Fog Index. To calculate the Fog Index of a passage at least about 100 words long:

(a) Find the mean number of words per sentence in the passage (i.e.,

TABLE 7. *Clues to nounification*

Clue	"Nounified" sentence or phrase	Shorter, livelier version ("reverbified")
"the __ of"	after the passage of the law	after the law was passed
	through the use of this system.	by using this system
	The analysis of the curve is simple.	Analyzing the curve is simple.
	the greatest problem in the development of the spacecraft	the greatest problem in developing the spacecraft
Other prepositional phrases	more evidence for the existence of black holes	more evidence that black holes exist
	under the direction of	directed by
"__tion" or "__sion"	(the above example)	
	A 30% weight reduction occurred.	Weight decreased 30%.
	results in their collision	makes them collide
	upon completion of the project	upon completing the project
	(the below example)	
"There is"	There is a wide variation in mortality.	Mortality varies widely.
	There will be some heat flow.	Some heat will flow.
Adjectives plus forms of "to be"	are dependent on	depend on
	are different	differ
	is fearful that	fears that
Verbs such as:		
"deliver"	delivered a lecture	lectured
"give"	give exposure	expose
"have"	have effects on	affect
"make"	make contributions to science	contribute to science
"produce"	produce relief of symptoms	relieve symptoms
"provide"	provide help to	help

the number of words divided by the number of sentences). Independent clauses separated by colons, semicolons, or commas count as full sentences.

(b) Determine what percentage of the words are "difficult." In general, words three syllables or longer fall into this category. However, exclude (1) proper names, (2) verb forms made three syllables by adding -ed or -es (e.g., "reacted" or "confuses"), and (3) combinations of short, easy words (e.g., "birdwatcher" or "turntable").

(c) Add quantities (a) and (b); then multiply by 0.4.[43]

Carbon
~~Although~~ ~~carbon~~ dioxide ∧ makes up only .03 percent
of the earth's
∧atmosphere ~~of the earth, present at a concentration of~~
However, it may play a
~~about .03 percent~~ by volume ∧, ~~it plays a possibly~~ critical
earth's
role in controlling the ∧ climate ~~of the earth~~ because it
thus
absorbs radiant energy at infrared wavelengths. Heat ∧
could alter
trapped ~~in this way,~~ ∧ ~~has a large potential for altering~~
the world climate substantially. ~~And quite apart from~~
The
~~possible effects on the climate,~~ ~~the~~ carbon dioxide in
provides
the atmosphere also ∧ ~~plays a critical role as the source~~
green plants fix
~~of~~ the carbon that ∧ ~~is fixed~~ in photosynthesis, ~~by green~~
Thus it serves as
~~plants and therefore provides~~ ∧ the basis for all plant and
animal life.

Mankind therefore faces a historic dilemma. The
~~human~~ activities ~~that are~~ increasing the carbon dioxide
warm
content of the atmosphere promise to ∧ ~~bring a general~~
~~warming of~~ the climate over the next several decades.

Calculations	Version	
	Original	Revised
Number of words	138	96
Number of sentences	5	7
Number of "hard" words	25	17
(a) Average sentence length (words)	27.6	13.7
(b) Percent hard words	18.1	17.7
Fog index = 0.4(a + b)	18.3	12.6

FIGURE 10. *Calculating and reducing the Fog Index. (Source of original passage: Woodwell GM. The carbon dioxide question.* Scientific American *1978 January:34–43.)*

The resulting quantity, or Fog Index, is the grade in school to which a passage's difficulty corresponds. For example, an average ninth grader should be able to read a passage with a Fog Index of 9 fairly easily; most high school graduates should find material with a Fog Index of 12 readable; and writing with a Fog Index of 25 is well into the postdoctoral range. The typical comic strip is at a level of 6, *Ladies' Home Journal* at 8, *Reader's Digest* at 10, *Time* and *Newsweek* at 11, and *The Atlantic Monthly* at 12; virtually no popular magazine scores higher than 12.

Figure 10 illustrates how to calculate the Fog Index and how, largely through principles presented in this chapter, to reduce it. In its original form, the sample shown has a Fog Index of 18.3—perhaps appropriate for *Scientific American's* highly educated readership, but unsuited for a broader audience. The revised version rates a much more comfortable 12.6.

To aid in deciding when to shorten words and sentences, you may well wish to calculate the Fog Index or a similar indicator (or to use word processing equipment that does so automatically). In particular, consider estimating readability of those passages that you somehow sense will bog readers down. If you have drafted a long article, perhaps sample a few sections and test them for readability.

Scores such as the Fog Index do have limitations. Although they suggest how easy a passage is to read, they cannot assess overall clarity. For instance, they do not indicate whether writing is logical. Indeed, a string of brief nonsense sentences could have a low Fog Index but would be impossible to understand.

"I've made this letter longer than usual," Pascal once apologized, "because I lacked the time to make it short." Communicating crisply can take considerable work. But in general, the more effort you put into expressing yourself simply, the less effort the audience must devote to understanding what you say. In presenting science to the public, trimness rather than ponderosity is a mark of success.

Chapter 15

Writing Effectively:
Other Advice

"Easy writing's curst hard reading" stated playwright Richard Sheridan. Such is certainly true of popular science writing, in which hard reading is at best poorly understood and at worst unread. This chapter therefore discusses several aspects of writing in which more trouble for the author can mean less for the audience.

Including the Human and the Concrete

Effective popular writing usually abounds with the human and the concrete. Yet the subject matter of science is often impersonal and abstract. This dilemma is milder than it might appear at first, for science includes people, processes, and products.

You can introduce the human element in various ways. One is by presenting material wholly or in part through human eyes, either your own or those of someone interviewed. Another is by including quotations. A third is by saying how an item affects, or could affect, people, especially those in the audience.

As for being concrete, using examples and analogies helps. So does presenting details that conjure up vivid sensory images: Just how, and like what, does an item look, sound, feel, smell, and perhaps taste? Also, try to avoid abstract terms. For instance, use words like "woman," "pilot," and "patient" rather than "individual"; "truck," "schoolbus," and "sportscar" rather than "vehicle"; and "lab," "college," and "gym" rather than "facility."

An excellent way to combine the human and the concrete is to tell a story, either as an entire article or as part of it. Doing so not only incorporates people and lets you present a specific case. It also helps in structuring the piece. And it may enable you to introduce such attention-

Chapter 7. THE STRUCTURE OF THE NUCLEUS OF THE ATOM
    "What?" exclaimed Roger, as Karen rolled over on the bed and rested her warm
body against his. "I know some nuclei are spherical and some are ellipsoidal,
but where did you find out that some fluctuate in between?"
    Karen pursed her lips. "They've been observed with a short-wavelength
probe . . ."

© 1975 Sidney Harris
American Scientist magazine

riveting elements as competition and suspense. Little wonder that some
of the most popular books about science have taken the form of tales.

Planting Guideposts

Various guideposts can help keep readers from losing their way. The
most subtle and pervasive are transitional words and phrases. These
unobtrusive items (such as "in addition," "thus," and "these") help
clarify the relationships among ideas.

Headings (such as "Planting Guideposts" above) also orient an audience. Depending on their purpose, they can be a single word ("Guideposts"), a phrase ("Planting Guideposts"), an entire sentence ("Use
guideposts to orient your readers."), or a question ("What guideposts
can you use?").

Another way to guide readers is by using underlining, italics or other special typefaces, or various sizes of type. These typographical tactics distinguish items of different importance (e.g., key terms from other material) and of different sort (e.g., case accounts from commentary).

A fourth device is to set off lists from the rest of the text. To do so, you can number items, place them one under the other (perhaps set off by bullets, as was done on pages 82 and 92), or both.

Granted, the presence of too many guideposts can distract an audience. Used in moderation, however, guideposts aid in various ways. Not only do they orient the public as it reads. They also make writing easy to preview and review. By varying the appearance of the text, they can make a document look inviting as well.

Keeping Antecedents Clear

Writing clearly demands more than adequate guideposts, short words, and short sentences. For instance, the antecedents of words such as "this," "it," and "which" must be immediately evident.

Unless followed by nouns, terms such as "this" and "these" tend to be ambiguous. Consider this pair of sentences: "Chemists hope to devise a protocol for synthesizing the compound. This could be very difficult." Just what, readers may wonder, is likely to be difficult—devising the synthesis, doing it, or both? Putting a noun such as "search" or "protocol" after the "this" removes the uncertainty.

Be sure that the antecedents of such words as "it" and "they" likewise are clear. For example, a student wrote: "Although human encounters with black holes have been limited to science fiction, there is evidence that they actually exist." Here the casual reader may think at first that "they" means "encounters" rather than "black holes."

Also, try to place words such as "which" and "that" next to the terms to which they refer. Otherwise, you may produce a sentence such as the science writer's lament: "I saw a story . . . about the world's first testicle transplant, which somehow our bureau managed to condense into 200 words and let die."[75]

Finally, if you include words of comparison, be sure to make clear what you are stacking up against what. Rather than merely calling something louder, tougher, or faster, also state "than what." And in addition to terming it "most effective" or "best," answer "among which?"

Unstacking Nouns

Scientists not only use many nouns; they also stack them one after another. For the public, considerable ambiguity can result. Does the designation "energetic materials chemist" mean a hard-working chemist dealing with materials or, as is actually so, a chemist dealing with energetic materials? Does "uncomplicated anxiety relief" mean simple relief of anxiety or relief of simple anxiety? Just what is a "third party payor designation"? Unstacking nouns and connecting them with words that show their relationship can keep the public from having to guess the answers to questions such as these.

Using Sex-Neutral Language

Not every nuclear engineer, astronomy buff, or biology professor is male. Nor is every nurse or secretary female. Yet the English language—which lacks a sex-neutral pronoun for a generic individual and traditionally has used "he" to mean "he or she"—makes it all too easy to perpetuate such stereotypes.

Repeatedly saying "he or she" and "his or her" avoids this trap, but it does so awkwardly. Fortunately, more graceful and less blatant tactics exist. The four pairs of sentences below illustrate some of the main ways to convert language into less sex-biased forms.

The typical science writer gets story ideas from many sources. For example, he scans scientific journals, attends conferences . . .

Typical science writers get story ideas from many sources. For example, they scan scientific journals, attend conferences . . .

The scientist wishing to write for the public should proceed as follows. First, he should analyze his audience.

As a scientist wishing to write for the public, you should proceed as follows. First, analyze your audience.

After the social worker sees a client, she files her report.

After seeing a client, the social worker files a report.

Usually, the student with a working wife will have less debt at the end of his graduate school years.

Usually, the student with an employed spouse will have less debt at the end of graduate school.

Keeping the Sticklers Happy

Is it acceptable to say "hopefully" instead of "I hope" or "it is to be hoped"? Is it O.K. to split an infinitive (e.g., to say "to quickly start treatment" rather than "to start treatment quickly")? And is it all right to use "which" instead of "that" in a restrictive clause (e.g., to speak of "the agency which funds our research" instead of "the agency that funds our research")?

Usually, choosing the former alternative will not compromise clarity or reading ease. But doing so may well annoy those readers who look conservatively on how language should be used. Thus, some members of the audience may fume about how uneducated you seem, rather than concentrating on what you say.

To avoid distressing and thereby distracting the sticklers, try to follow the conventional rules when feasible. (For a quick rundown on traditional usage and much more, Strunk and White[82] is excellent.) Such caution can help keep readers satisfied and attentive. It may also aid in winning the respect of, and assignments from, editors.

Avoiding a Distracting Style

Straying from traditional usage is not all that can distract an audience. Launching too many verbal fireworks also can. As emphasized throughout this book, such devices as analogies, examples, quotations, and lively wording help readers enjoy and understand what you say. Such items should enhance your message, not compete with it. Thus, use these devices subtly and sparingly. Ideally, those who read or hear a piece on science should come away thinking "That was so clear and interesting!" not "Wasn't that cleverly written!"

Refraining from the Ridiculous

Check your words, and perhaps have others review them, to be sure that they are not likely to be misinterpreted as meaning anything ludicrous. Be especially alert for items such as the following:

> Lichtenberg, the first non-NASA scientist to fly on an American space mission, wrote his Ph.D. thesis on an acceleration sled similar to the one that will test motion sickness susceptibility before and after these Spacelab flights. (The guy must really have the right stuff. Most of us feel ill when writing a dissertation at a desk.)

> Roberts, the father of four children and an intellectual, advocates teaching science more rigorously in high school. (Is Roberts or some non-child offspring of his the intellectual?)

> Dr. Jones studied mainly adults with normal hearing. She also

"Oh nothing much. Just experimenting with some rabbits."
Drawing by Levin; © 1980
The New Yorker Magazine, Inc.

included subjects with hearing impairments, children, and those with ringing in the ear. (I know that some consider children an affliction, but . . .)

The equipment has an imaging magnet of a size to accommodate a human head as well as small research animals. (I would be nervous enough during a brain scan without having beasties huddled next to my skull.)

Unless aiming for hilarity, also beware of combining too many images. The passage below illustrates the pitfalls of adding image-laden text to a quotation that already mixes metaphors:

> "We thought we hit a home run," explains Basil J. Whiting Jr., the Labor Dept.'s deputy assistant secretary for OSHA, "but they're telling us to go back to first base. We've been cast adrift without any clear guideposts." Indeed, the court's 5-4 vote leaves wide open many long-simmering issues that are at the heart of industry's complaints against OSHA.[18]

Finally, be alert for silly-sounding combinations of words. "For example, for a sample . . ." and "overcoming the foregoing shortcomings" may be right for a patter song or parody, but they hardly enhance a popular article on science.

Checking and Editing

Just as writing is not the first step in producing a story, neither is it the last. After drafting a manuscript come checking and editing. Like scientific papers, popular writing often needs revising again and again.

In checking, see that all names, statistics, quotations, and other items are accurate. To do so, consult your original notes and references and, if needed, other written and human sources. For future reference, record sources in the margin of the manuscript, opposite the facts obtained from them.

If time permits, set the manuscript aside for a while before editing it, so that you can approach it freshly. Then read it and ask yourself questions such as those in Table 8. Perhaps also test the draft on readers

TABLE 8. *Twenty questions to ask when editing popular science writing*

1. Does the format seem suited to the item's function? Does it meet all requirements set by editors and others?
2. Are the content and style appropriate for the audience's background, interests, and needs?
3. Does any information seem to be missing?
4. Does any information seem to be superfluous?
5. Does the beginning of the item orient and interest the reader? Does the end bring the account to a satisfying close? In between, is the material logically organized?
6. Does the item have plenty of human interest, if appropriate?
7. If the item is based in part on interviews, does it use quotations effectively?
8. Are enough major guideposts present to orient the reader?
9. Does the item flow smoothly? If not, how could transitions be improved?
10. Is every sentence worded clearly? In particular, are the antecedents of pronouns clear?
11. Is the writing crisp rather than "foggy"? Are there places in which to remove needless words, substitute brief words and phrases for long ones, and otherwise shorten words and sentences?
12. Does the item contain techical terms? Should everyday language replace them? Should they be defined (or be defined better)?
13. Are statistics, if any, presented in a form that readers can understand?
14. Are spelling, grammar, and usage correct?
15. Is too much passive voice present?
16. Is the language sex-neutral?
17. Does anything sound odd or ridiculous?
18. If the item has visuals, are they effective? If it lacks visuals, should it have them?
19. Does anything about the item make you uneasy? If so, what could remedy the problem?
20. Is the item one that you would want to read (or see or hear)? If not, what could make it so?

resembling the intended audience. Then, based on what you think and hear, revise the manuscript.

Writing that follows principles such as those in this chapter and the last is likely to be blessed easy reading. In fact, others trying to present science to the public may well call on its author as an editor.

Chapter 16

Preparing a Book Review

"To me, criticizing implies celebrating and loving," commented Steven J. Marcus, then the managing editor of *Technology Review*, during the MIT lecture series "Celebrating and Criticizing Science: The Responsibility of Popular Writers."[55] In book reviewing, the link between criticizing and celebrating is especially close. If you value popular works on science, want them to be widely recognized and used, and hope to make them better and more plentiful, consider writing reviews.

For brevity, this chapter often will call the works reviewed "books" and the audiences for both the works and the reviews "readers." Most of its content, however, pertains to reviews of broadcasts and films as well as of written items. Likewise, it applies to reviews prepared for the broadcast as well as the print media.

Appreciating the Value of Reviewing

In various ways, reviews can increase the public's awareness and understanding of science. Of course, they can direct readers to sound, interesting, well-prepared works. In addition, they can amplify such works' impact by presenting some of the content from them, as the example in Figure 11 does. (How often have you heard or said: "No, I haven't had a chance to read the book, but I'm familiar with it from a review."?) In fact, the audience for reviews may be larger than that which reads news articles or features on science. In addition, a well-reasoned, clearly written review can help readers learn to consider science critically.

If you belong to the main group for which this book is intended, you are well suited to review popular works on science. As a scientist, you can determine whether content in your field is sound. If this book has achieved its goals, you know and can apply principles of presenting

LIFE IN MOVING FLUIDS: THE PHYSICAL BIOLOGY OF FLOW, by Steven Vogel. Willard Grant Press, 20 Providence Street, Boston, Mass. 02116 ($18). Not even a strong rinsing flow is as effective for cleaning a surface as any swipe of a dishcloth. Dust accumulates on whirring fan blades, pipes scale up and thicken instead of wearing smooth and erosion by streaming water is wonderfully enhanced by a little sand. In fact, right at the solid surface there is stillness. Fluids simply do not slip over solid surfaces, whether they are rough or smooth, greasy or clean. The boundary layer always holds a velocity gradient. That was the great contribution of the early 20th century to fluid mechanics. Ludwig Prandtl first recognized that even in turbulent flow the thin layer at the boundary remembers that real fluids are viscous; therein was the solution to the paradoxes of classical hydrodynamics. From that success the eddies flowed, both in the air and in the theory, to bring lift, drag and flight under the laws of Leonhard Euler and Isaac Newton.

Life has known it all along, of course. Organisms dwell in fluid flow, in wind or water, beating surf or the stagnant air at the grass roots. Much depends on what passes the boundary layer and how fast: water, heat, edible particulates and the propagules of many life forms. Diffusion is all very well for bacteria, but once organisms are big enough they can walk and swim. Along the scale, however, something must help the tiny spore and pollen particles to gain access to "the free transit system in the sky."

For doubters there is a fine control study. Club-moss spores were allowed to settle on paper models of leaf surfaces. Then they were blown free in the wind tunnel. It took storm winds to lift the tiny spores off surfaces with a roughness comparable to the spore diameter. Any little elevation helps greatly; in a modest breeze even a millimeter is a real winner. Many are the inventions of life. A barley smut manages to stiffen the infected heads so that they do not bend in the wind, and the spores float free. The famous ballistic fungus *Pilobolus* is not the only gun carrier; its optimization is a delicate trade-off. The gun fires a small spore case, bigger than any spore. Small spores would not go as far, but they would fall out more slowly. Puffballs, poppies, sage, lichens and mistletoes have their own little catapults, some with a power assist from raindrops or the wind itself.

Out on the Pacific coast (perhaps related to the sophistication of the aerospace industry?) there is a colonial kind of sand dollar. It lives on edge in the tidal currents. The lift its cambered form develops is substantial, but with its edge buried firmly in the sand the creature does not move at all. The lift is horizontal, plainly of little interest as a force. The streaming circulation measured by the lift, however, brings the passing particles closer to the feeding appendages on the surface. These echinoderms always cluster, all parallel on edge, a multiwinged arrangement of hydrofoils with their spacing adjusted over a factor of 10 to the surge velocity of the local currents.

"Fluid flow is not currently in the mainstream of biology," writes this author in his personal, iconoclastic way. He has nonetheless written a model of a book, halfway between an introductory physics of flow and an overall sketch of many problems of life in the stream. Whether he is analyzing the simple equations he uses or telling how to measure the drag of an obstacle by indirect means, his sense of aptness and physical understanding comes first. "I once measured the lift of a fixed fly wing by first measuring the airspeed at each of a series of points behind it; then for each point I centered the wake of a tiny wire on the airspeed transducer to get the local wind direction. From these data it was a simple matter to calculate the downward component of momentum."

Perhaps a third of the book is that artful an introduction to the physics of flow—supersonics and hot gases apart—

FIGURE 11. *From a review in "Books" by Philip Morrison. Copyright © 1982 by Scientific American, Inc. All rights reserved.*

as life encounters it. The Reynolds number is a powerful motif, beautifully modulated; the range examined finds this key dimensionless index from 10^{-7} deep in the tiny filter sieves of the plankton up to 10^8 and more for the free-swimming whale. (How whales can filter feed as well as they do has not yet been studied.) Vogel tells of the drag of trees and weeds and insects, of the life in velocity gradients mentioned above, of flow through pipes, of lift in soaring, gliding, flapping and just spinning down quietly in the autumn twilight. Flow through pipes is treated in sketch outline; it is the most complex, if medically the most important: "I just don't want

to dwell," Vogel concedes wisely, "on the nonsteady flow of a non-isotropic, non-Newtonian fluid through nonrigid pipes. Something of the order of 10^4 papers appear on the subject of blood flow each year."

For biologists who want to come to the beginnings of a quantitative understanding of a wide variety of adaptations, for general readers who want to see how fluid mechanics works in a varied and often surprising context, for lecturers and students who want to read a text that seems as personal as a confidential chat, this book, full of data, rich in up-to-date and well-appraised references, is a first-class opportunity.

FIGURE 11. *(Continued)*

science to the public. Thus, you can tell how accurate a popular work is, judge how appropriate it is for its target audience, and prepare an effective review for a lay readership.

Reviews can be among the most convenient types of popular writing for scientists to do. In preparing a review, one often need consult little or nothing more than the work at hand and one's own expertise. Also, most reviews are relatively brief.

In addition, reviews are a fine way to begin writing for the public. An editor hesitant to commission a full-length feature article may let you try a book review. Then, if all goes well, other assignments may follow. Furthermore, enclosing copies of your book review(s) can help convince other editors of your writing skill, and editors reading your reviews may contact you on their own.

Determining Where to Publish Reviews

Many popular publications, as well as some broadcast programs, run reviews of popular works on science or could do so. As one scientist experienced in publishing observes, such reviews might well be more common if editors knew of more individuals qualified to write them. If you inform the appropriate editor that you are available and perhaps mention a work that you would like to critique, you may be invited to prepare a review.

If you would like to prepare brief reviews, consider volunteering to write for *Science Books & Films* (1101 Vermont Avenue, N.W., 10th Floor, Washington, D.C. 20005). Published five times per year by the

American Association for the Advancement of Science, this magazine consists almost solely of 200-word reviews (such as the example in Figure 12) of books, films, videocassettes, and filmstrips in the various sciences. Works reviewed include those for students, the general public, and professionals; nearly all of the reviewers are scientists and educators. Because many librarians and teachers consult *Science Books & Films*, reviews in it can strongly affect what the public reads and views.

However, most reviews for the public appear in, or in supplements to, publications containing much more than such critiques. Both daily and weekly newspapers sometimes feature book reviews by writers not on their staffs. So do various magazines, including some specializing in science and technology. If you are interested in reviewing books or other items for a newspaper or magazine, contact the appropriate editor. Also, consider preparing reviews for the broadcast media.

Writing reviews is rarely a way to get rich. You may receive a stipend (most likely a small one), be reimbursed for or given the book, receive at least one free copy of the publication in which your review appears,

---

THE SHARKS. National Geographic Society, Educational Services, 17th & M Sts., NW, Washington, DC 20036; 1982. 59 min. Color. $595 (16mm); $545 (video); $42 (rental). o.n. 05405 (16mm); 05406 (video).

JH–P, GA ★★   Despite a sensational opening sequence of shark attack, this informative and exquisitely photographed film eventually succeeds in developing the more realistic views that sharks are a varied group of fishes and that only a fraction of them are dangerous to humans. The tone of appreciation for and conservation of shark diversity is diminished only in the film's failure to note that certain shark populations are threatened with decline as a result of intensified fishing for them in various parts of the world. Viewers are treated to some spectacular footage—for example, basking shark, whale shark, great white shark, and blue shark—and learn that sharks are equipped with an extraordinary range of sensory capabilities. Various scientific investigations are highlighted, including those of Eugenie Clark, the scientific advisor for the film. She is featured in several segments and is an appealing blend of enthusiasm, curiosity, and scientific authority. Clark is particularly effective in stressing the beauty and grace of sharks and the relative harmlessness of most of them. The film ends on a note that offsets the ominous beginning—a shark is photographed as it hatches from an eggcase that is attached to the ocean bottom. This final sequence represents the potential for the harmonious coexistence of sharks and humans. The quality of the film is enhanced by the judicious use of background music and by clear and precise narration. This film has something to offer to wide variety of audiences.—*Michael H. Horn, California State Univ., Fullerton, CA*

FIGURE 12. *A film review. Reprinted with permission of AAAS,* Science Books & Films, May/June 1983.

or get some combination of the above. The main rewards are likely to be seeing (or hearing) your work and knowing that you are helping to inform the public about science.

Deciding What to Review

As implied above, editors may propose works to review, and you can suggest items yourself. One source of information on forthcoming books is the magazine *Publishers Weekly.* Also, if specific companies often publish material in your field, you may ask them to send you announcement of new books. Such advance notice can be important, for most major magazines and newspapers review books near the time of publication. Some outlets for reviews, however, have more flexible timetables; in seeking books to review for them, browsing in bookstores and noting reviews by others can also be of aid.

One difficult question is whether to bother reviewing a work that you consider poor. If such an item seems destined for a quick and quiet death, the best approach may be not to interfere; save your words for when you want to change something. If, however, the work is being heavily advertised, widely praised, or both, you may be serving the public by pointing out its shortcomings.

Another dilemma arises if you are asked to review a work by a rival or a friend. If you feel that you cannot be objective, or even that others will doubt your objectivity, the assignment might best be refused. If you do write the review, perhaps include a phrase (for example, "with whom I have long disagreed" or "one of the colleagues whom I respect most") that will help alert readers to your orientation.

Preparing to Write the Review

Book reviewing, like other writing, is largely a matter of preparation. A basic step, as in other assignments, is to obtain clear instructions: when to submit the review, what sorts of information to include, and what format to follow. If you are not familiar with the publication requesting the review, become so. Ask the editor about its readership and emphasis. Peruse several issues, and note their subject matter, slant, and style. Look especially carefully at the content and construction of book reviews published previously.

One obvious bit of advice, but one that book review editors note is more than occasionally violated (sometimes with embarrassing results), is to read the book carefully and without preconceptions. In addition to reading the book thoroughly, perhaps skim it beforehand, afterward, or both in order to obtain an overview. To avoid groping for the passage that you wanted to quote or the statistic that you wanted to cite, take notes, either in the book or separately. Consider jotting down not only salient points and important passages from the work but also observations that occur to you—and even specific words, phrases, and sentences to use in the review.

Talking to others about the book before you write the review may help you to formulate your ideas and to find out what the public wants to know. "Use your family and friends as sounding boards," freelance book reviewer Jan Frazer advises. "Tell them about a new book you've finished reading and watch their faces for the interest spark."[35]

Writing the Review

Book reviews vary from mere synopses ("book reports"), to combinations of summary and evaluation, to essays in which the book is only a launching point for a discourse on the reviewer's own ideas. The suggestions below pertain mainly to the second, and principal, category. Figures 11 and 12—examples of popular reviews of different lengths, for different segments of the public, and of works in different media— illustrate various elements of good reviews.

Of course, a review should note the main sorts of information and ideas in a work. Ideally, it should actually present some of that content, for doing so not only helps the audience decide whether to read the book but also educates the public on its own. In addition, a review for the public should place the work in scientific and general context. "Give the so-what," Barbara Seeber, book review editor of *Science 83*, emphasizes. "The scientist doesn't need the so-what. The average reader does."

The review should also address the item's strengths and weaknesses as a work for the public. Thus, a reviewer should assess more than whether the content is valid. Barbara Walthall, editor of *Science Books & Films*, notes the importance of considering whether a work clearly distinguishes fact from theory. Other questions include whether the amount of detail is appropriate for the intended audience and whether the style is clear and interesting. Presenting evidence, such as quotations, to support the evaluation not only strengthens a critique but also helps to illustrate sound scientific thinking.

A work's visuals and physical qualities can also be worth discussing in a review. Photographs and diagrams are sometimes crucial to a popular science book; you may well want to comment on them (or on the lack thereof). If the typography or bookmaking is exceptional in any way (be it favorable or unfavorable), perhaps mention that, too. Also consider the relationship between a book's function and its form; for instance, is a field manual sturdy and of a convenient size?

Another item to mention is the audience for whom the work is appropriate. In addition, comparing a book with others (How does this home medical encyclopedia rank among its competitors? Is John McPhee's latest volume as good as his other ones? If you already have the first edition of this guide, is buying the second worthwhile?) may enhance a review's usefulness.

The reviewer's main role is to evaluate a book evenhandedly and independently; beware of temptations that can interfere with doing so. Take care not to show off erudition for its own sake; the objective is to demystify science, not to make it seem obscure. Avoid needless sarcasm, wittiness that compromises accuracy, and nitpicking. Also, if others have reviewed the same book, be cautious about reading their critiques before at least composing a draft.

A book review should itself exemplify good science writing. The content and style should suit the audience's background, interests, and needs; the language should be simple, concise, and concrete; and the writing should follow other principles of popular science communication. Thereby a critique will achieve its full potential to celebrate presenting science to the public.

Chapter 17

Making an Oral Presentation

Terror, confusion, distaste, embarrassment, and fatigue. Such are the feelings that many of us associate with giving a speech. For much of the scientific community, delivering papers to peers is an unwelcome task— and public speaking is downright threatening.

Yet oral presentations can be among the most effective and satisfying routes for presenting science to the public. First of all, they have human interest built in: the scientist. Second, they can draw on both hearing and sight, and sometimes other senses, both directly and through a wide range of audiovisual aids. Third, unlike writing, speaking permits immediate two-way interchange.

Whether public speaking proves a punishment or a reward depends largely on how one approaches it. As John B. Bennett, who has lectured extensively on oral communication at the Harvard Business School and elsewhere, notes, "Take care with your preparation, take pains with your rehearsal, and to a great extent the presentation will take care of itself."

Defining the Audience and the Task

In public speaking as in other communication, defining the audience and the task is basic to success. Table 9 presents twelve items to consider when you will be giving a speech. This checklist also can be used in instructing speakers whom you host.

Making a Presentation Clear and Interesting

When he was secretary to The American Physical Society, Karl K. Darrow suggested to its members the following demonstration:

TABLE 9. *Defining the audience and the task: Twelve basic items to check when planning a speech*

1. What are the function and composition of the group that you will address? How much will the audience know about your field?
2. At what event will you be speaking? Does it have a theme with which to link your remarks?
3. Will the presentation be part of a series? If so, should you coordinate your presentation with those of the other speakers?
4. Why will group members be attending the event—mainly to hear you or for some other reason? What is their attitude likely to be toward you and your field?
5. Are any matters touchy? For instance, are there any topics, protocols, or people to treat especially sensitively?
6. What are you supposed to discuss? How much leeway do you have regarding the topic?
7. What are the objectives of your talk? For example, should you offer mainly general enlightenment, practical advice, entertainment, or something else?
8. How long should the presentation be, and will a question-and-answer period follow?
9. How large will the group be, and in what size and sort of room will it be meeting?
10. What audiovisual equipment will be available? If you may be requesting equipment, how and by when should you do so?
11. Will adequate voice amplification be available?
12. Should you provide anything (e.g., your résumé or curriculum vitae, a copy of your remarks, or handouts) in advance or at the time of the speech?

> Choose an article in *The Physical Review*; let it be in your own field if you will, lest the result of the experiment be too frightful. . . . [R]ead the article—but read it according to the following prescriptions. Read straight through from beginning to end at the rate of 160 to 180 words per minute. Never stop to think over anything, not even for five seconds. Never turn back. . . . Never look at an illustration until you get to the place where it is mentioned in the text; and when you get to that place, look at the illustration for ten or fifteen seconds and never look at it again.[20]

Imagine subjecting a member of the public to the above punishment, and some of the differences between writing a scientific paper and giving a popular presentation should become evident.

When readers find a written passage unclear, they can review it again (and again) at any pace until they understand. If their attention lapses, they can return to the item at any time and not have missed a bit. However, when part of a speech confuses audience members or lets their attention flag, they may well remain bewildered and stop listening. Thus, in oral presentations even more than in other forms of communication to the public, special attention is necessary to keeping the material *clear* and *interesting*.

The form of a speech, of course, is essential to *clarity*. Thus, in preparing a presentation, consider the guidelines below:

- Keep the speech's overall structure simple. For example, trace a project from start to finish, compare options, or present examples illustrating a principle. To help listeners follow along, promptly indicate what pattern the speech will take.
- Limit the number of points made. Also, present major ideas far enough apart so that the audience has time to grasp each; intersperse supporting material such as examples and anecdotes.
- Try to keep sections of a speech independent enough so that listeners who fail to understand or remember one part can nevertheless gain something from the rest.
- Repeat major points immediately to make sure that listeners grasp them. Perhaps also summarize at the ends of sections and of the entire speech.
- Make the relationships among ideas clear. For example, include transitional phrases (e.g., "the second reason," "in other words," and "to summarize") to help the audience to keep its place.

In maintaining clarity, one advantage of oral over written communication is the continual feedback from the audience. Do listeners look bewildered? Does attention seem to be flagging? Do questions from the audience suggest that the message has not come across? If so, perhaps clarify what you said.

As for snagging and holding the audience's *interest*, various tactics can be of aid. To nearly any popular science communication, the human sort of interest literally adds life. Whereas this element can be hard to include in writing, oral presentations have intrinsic human interest, for a person delivers them. To help keep the audience listening, take advantage of its interest in you. Do not hesitate to show expression and personality; if appropriate, talk about your own experiences (doing so, by the way, offers a fine chance to portray the process of science). Perhaps tell anecdotes about other people as well.

Relating what you say to listeners' preexisting knowledge and concerns also draws interest. So do such devices as making part or all of the material into a narrative and injecting tasteful, relevant humor.

Giving listeners an active role also helps to retain their interest, and it may help them to retain what you say. If the audience is small and the setting is informal, options include asking listeners questions, inviting them to interrupt with comments, involving them in demonstrations, and letting them try out apparatus. Even if a group is large, you can give it a sense of participation through such approaches as rhetorical questions, thought experiments, and self-quizzes.

Finally, aesthetic elements add audience appeal. One need not be a majestic orator to present science effectively to the public. In fact, flashiness can distract listeners from the message to be conveyed. In general, a simple approach is best. Direct, vigorous wording; uncluttered audiovisuals; and a natural, friendly speaking style all help to make a presentation pleasant and interesting.

Preparing and Using Notes

As those of us who attend scientific conferences know all too well, hearing a paper read verbatim is often deadly dull. Many a speaker stares down at the manuscript the entire time and mutters a mile a minute in a monotone. Often the paper itself is written in a format that would be fine in print but is ineffective orally. Such a presentation is trying enough for the scientist intent on understanding it. The layperson confronted with such speaking might well give up, convinced that scientists are indeed bores.

Reciting a memorized speech can be at least as bad. If you have memorized your speech, you probably are concentrating on not forgetting anything, and you probably have reviewed your words so much that you are thoroughly sick of them. Little wonder that you may sound like an automaton. Also, should you lose your place, you may not know what to say next.

Likewise, speaking without preparation is usually disastrous. As Mark Twain observed, "It takes three weeks to prepare a good ad-lib speech."[66] Few of us can spontaneously deliver an effective talk even to our closest peers; presenting science in a form that the public will appreciate and understand takes particular planning and work.

In general, the best approach is a compromise: *use notes.* Glancing at them from time to time, speak conversationally. Thus, combine the careful planning necessary for nearly any good science communication and the spontaneous style that makes human talk warm, lively, and interesting. In technical terms, give what is called an extemporaneous speech.

How to prepare notes is a matter of choice. Some experts favor producing a manuscript and then transforming it into notes; a rule of thumb is that one double-spaced page equals about two to two and one-half minutes of talk. Others recommend working all along from an outline or a set of file cards. Your ultimate set of notes should likewise take the form most convenient for you. Many speakers prefer file cards. Others use sheets of paper; extra-large type and triple spacing can aid visibility. Whatever the format, make an extra copy of the notes in case one is mislaid. "Also, be sure to number your file cards," says one scientist who learned the hard way.

Notes should contain both less and more than a speech's content. They should not state the entire text; rather, they should include such items as ideas, phrasings, and statistics, in order to help ensure that the talk is complete, effective, and accurate. To be most useful, they should also indicate such matters as when to introduce audiovisuals, where to pause or to change inflection, what part of the presentation to reach by when, and which passages to omit if time falls short. For clarity, consider marking such stage directions in a different color than the body of your notes.

Of course, rehearse the presentation before giving it. Doing so helps you to devise effective phrasings, lets you adjust the timing (speeches often run slightly longer "on stage" than in rehearsal, so plan accordingly), and aids in establishing your confidence. Perhaps practice the speech for individuals resembling the intended audience; maybe audio- (or even video-) tape yourself. Beware, however, of rehearsing too much. Better a presentation with a minor rough spot or two than one of which you are too tired to impart the enthusiasm that brings oral communication to life.

Delivering the Presentation

From the time that you arrive at the lecture hall until you bid the audience goodby, various matters of style can help guarantee success.

Most of us experience "butterflies in the stomach" before we give a speech. "I regard that as a good thing," oral communication expert John B. Bennett states. "You have to psych yourself up." The rush of adrenaline that may be making your heart pound is probably giving you plenty of pep and making you extra alert. Moving around a bit can help you comfortably discharge any excess energy before you speak.

Then comes the presentation's opening. This crucial section has at least three goals: to present the speaker effectively from the start and create the desired atmosphere, to capture the audience's interest, and to announce the subject and structure of the speech. In achieving the first objective, showing confidence yet avoiding arrogance is the key. Let both the sound and the content of your words be friendly; perhaps thank the group for inviting you and include a phrase or two (for instance: "as Professor X, a graduate of this institution and one of this country's foremost engineering educators, once said . . ." or "in keeping with the theme of the YZ Society's fund drive for this year, I've decided to focus my remarks on . . .") from which the group can infer that you know and appreciate its activities, achievements, and goals. Be aware that an expert presenting science to the public may inadvertently sound condescending; one way to help counteract this tendency is to use words and manner indicating that you will *share* ideas with, not bestow them on, the audience.[1]

As for achieving the introduction's second goal of interesting the audience, many tactics are available. Unless you are a gifted storyteller, beware of starting with a joke. Devices that are easier for most speakers to use include reciting a relevant quotation, asking a rhetorical question, making a surprising statement, setting up suspense, and presenting an example. If the audience already cares about your topic, however, the most effective strategy may be merely to state the subject, perhaps briefly confirm its importance, and then proceed right ahead.

Whatever the sort of opening, it should not be so long as to bore or distract the audience. Rather, it should promptly fulfill its third function by noting the subject and structure of the speech. If you will be presenting a specific point of view, perhaps state it right away; should you feel, however, that the audience may be hostile to your opinion, perhaps build up to it gradually. The beginning is also the time to state ground rules—for example, whether audience members may interrupt the speech or whether they should save questions and comments until the end.

Scientists, especially those addressing the public for the first time, often "put tremendous effort into making a good lead but then start drifting into jargon," observes James Cornell of the Harvard-Smithsonian Center for Astrophysics, which sponsors a series of popular astronomy lectures. The principles of presenting science to the public should remain in play from a speech's beginning to its end.

Speaking style also is important in presenting science effectively. The main advice on delivery, though, is no advice: *be yourself.* Be yourself—but be yourself at your best. The following hints may aid in doing so:

- Show interest in your subject. If a speaker seems bored, why should listeners consider science (and scientists) anything but dull? The chance to put enthusiasm into your voice, to gesture, and to show expression on your face is a major plus of oral communication over print. Use it to attract the audience and to reinforce your words.
- Use timing to your advantage: to attract attention, to separate ideas, and to emphasize important points. For example, pause between elaborating on one item and introducing another. Likewise, consider expressing major conclusions slowly and deliberately.
- Maintain plenty of eye contact. Many speakers, especially those unaccustomed to addressing the public, are most comfortable mumbling into their notes and pretending that they are alone. However, looking at the listeners can help both them and you. Eye contact aids in keeping the audience's attention. It also lets you gauge the group's reactions so that you can adjust your remarks accordingly.
- Be alert for potentially distracting habits. Nervousness about public speaking can readily give rise to fidgeting. Awareness of this tendency will help in avoiding it.

Despite the importance of style, a presentation need not be flawless to be good—or even to be outstanding. As long as speakers maintain their composure, they can misspeak and then correct themselves, forget to mention details, contend with uncooperative audiovisuals, or worse, and still do fine jobs. Listeners may not even notice such problems. If some do, they will probably identify with the speakers for their fallibility and respect them for their poise.

A benefit of speaking to the public rather than writing for it is the chance to exchange ideas immediately. To make best use of this opportunity, anticipate likely questions and think of replies; perhaps have others quiz you in advance. But no matter how well prepared you are, questions that are difficult to handle can arise. If an inquiry seems irrelevant even once you ask that it be clarified, one tactic is to shift attention with a transition such as "That's a fascinating question, but as I see it a much more basic issue. . . ." Another strategy, useful also when a persistent questioner is hostile or is seeking information of little interest to the rest of the audience, is to suggest that the individual talk with you afterward.

Sometimes listeners have questions but no one wants to be the first to ask. Therefore, consider arranging for an audience member whom you know (for instance, your host) to begin the questioning. Perhaps even have such an individual ask you about a specific matter that you consider important but lacked time to discuss in your talk.

As for ending a speech, it need not be with a bang, but it should not be with a whimper. Beware of trailing off with a faint and apologetic, "Well, I suppose that's all I have to say. . . ." The tone should remain vigorous to the end. Perhaps conclude with something lively such as a quotation or anecdote; circling back to a device used in your introduction sometimes works especially well. If, however, your main goal has been to convey information and that is what the audience wanted, the best approach may be simply to summarize what you said. Also, if you have been trying to impart a "take-home message," restate it clearly and vividly, so that it will echo in your listeners' ears. And if you have enjoyed the chance to speak, consider saying so.

Using Audiovisuals

One of the best speakers I know uses audiovisuals galore. In fact, he often shows two or more slides at once. People today, he says, are accustomed to television and want lots to see. Also, he notes, audiovisuals help to clarify his material.

Another of my favorite speakers rarely uses audiovisuals. His careful planning and excellent delivery more than suffice in interesting and informing his audience, and his field does not lend itself easily to visuals. Why struggle with audiovisuals, he asks, when they will add little to, and may even distract from, his talks?

Whether to use audiovisuals (and, if so, which ones to use) depends on both subject and personal style. It also depends on the size of the audience and the setting for the speech. (Visiting a lecture room beforehand, if possible, can assist in planning audiovisuals.) Finally, it depends, of course, on what apparatus is available. One seemingly obvious but oft-neglected principle is to check that the desired equipment will be present and working; if in doubt, bring your own along, or at least take key components such as chalk, an extension cord, a pointer, and an extra projector bulb.

In considering audiovisuals, remember that they are audiovisual *aids.* Properly used, they clarify, reinforce, and supplement oral communication; they do not substitute for or compete with it. Thus, keep materials as simple as they can be and still convey their message: limit the number of points made in each, and exclude extraneous details. Also be sure to show visuals for long enough so that viewers can absorb the content, but do not present them so early or leave them on so long that they distract the audience. Third, do not rely on aids alone to convey the information; describe their content and discuss its significance.

Many different audiovisual aids are available to suit various needs and constraints. The *blackboard,* among the simplest and least expensive to use, is well suited for presenting brief messages and simple diagrams to small audiences. Its schoolroom connotations may be a plus in some situations but a minus in others. Speakers using blackboards should generally plan in advance what they will write and draw.

Flip charts are another way to present brief messages and simple diagrams to small groups. You can prepare them completely in advance, or you can sketch words and drawings in pencil and then trace the lines in felt-tipped marker at the relevant times in a speech. One option is to leave alternate pages blank, so that you need neither leave a chart in view once you are done with it nor show the next prematurely.[1] If you want to show more than one page at a time, you can tear off pages and tape them onto the wall.

Because of its convenience and versatility, the *overhead projector* is a favorite of many speakers. It is suitable for various sizes of settings, can be used in a normally lit room, and lets you face the audience. You can make the overlays before a speech (either with markers or on a copying machine), write or draw during the presentation, or both. If you have prepared an overlay in advance, you can cover it and reveal it a bit at a time. Also, you can pile overlays on top of each other to build up diagrams of increasing complexity.

Slides, which can be seen by nearly any size of audience, can lend great authenticity and immediacy; when you cannot bring the real thing into the lecture hall, slides can be an excellent second best. They also are a fine way to present drawings and cartoons. To be effective, however, slides must be appropriately made and used. In order to be legible, a typewritten slide should be double spaced and contain at most six to nine lines of text; each line should be no more than 45 characters long.[6] Diagrams presented on slides should be simple as well. Repeatedly switching lights off and on can distract an audience, but leaving the room dark too long between slides can lull viewers to sleep; therefore, try to group slides in one bunch or a very few. Plan far enough in advance, so that slides are ready and that functioning equipment is available.

If a topic has an auditory component (say, the call of an animal or the speech produced by a machine), *audiotapes* can help bring a talk to life. A relevant snatch of music or conversation can enhance a presentation, too. Obtaining the proper equipment will help ensure that listeners hear the material loud and clear.

Especially when movement is basic to a subject, a *videotape* or *movie* can both enliven and clarify a talk. Consider using part or all of a commercial film or tape, or of footage made in the course of research; if a sound track exists, think about whether to use it or to provide words of your own. Of course, be sure that the equipment is suited to the setting, so that the full audience can see (and, if appropriate, hear); to avoid lags during a talk, have everything set up in advance.

If the subject, type of audience, and setting permit, another option is to show *the real thing* or a *model* thereof. A related approach is to conduct a *demonstration*. At their best, such as in Michael Faraday's nineteenth-century discourse "The Chemical History of a Candle,"[32] demonstrations can combine drama and information most effectively. At their worst, they can be confusing, condescending, or mere flops. If you will indeed be giving a demonstration, not only prepare carefully to make it work; also plan how you can make your presentation work even if the demonstration does not.

Finally, especially if a presentation has a take-home message, consider giving listeners *handouts* to take home. For instance, distribute outlines, brochures, or reading lists before or after you speak. Any such items should, of course, follow the basic principles of popular science writing.

Oral presentation offers many chances to make science clear and interesting to the public. Plan carefully to take advantage of these opportunities, and most likely the wretched feelings usually associated with public speaking will give way to satisfaction and pride.

Chapter 18

Pursuing a Career
in Science Communication

Presenting science to the public can be important, feasible, and rewarding for many in science. A few scientists value this activity so highly, prove so adept at it, and find it so satisfying that they devote their careers totally or in part to it. If you enjoy informing the public about science, like dealing with a wider range of topics than is possible in most scientific posts, and wish to function independently yet meet many interesting people in your work, science communication may be a fine area for you.

Recognizing the Niches

Science communication contains a wide range of niches. Some of the most visible are in the mass (and semi-mass) media. Science-oriented positions for editors, reporters, and others exist at science, health, and technology magazines; at other periodicals (e.g., news, business, and women's magazines); at newspapers; and in the broadcast media. The number of such jobs seems to be growing, but so does the number of people interested in them. Individuals combining scientific backgrounds and communication skills are likely to compete well for the available posts. Science training is a particular asset in seeking employment at trade magazines, newsletters on science and technology, and other semi-technical publications.

Opportunities also exist at public information offices, where a scientist's presence can both enhance the media's coverage of science and aid in presenting science directly to the public. A scientist may either work full time for such a unit or combine work for it with other responsibilities such as administration, research, or technical writing or editing. Locations of PI offices dealing with science include universities, research institutes, government agencies, corporations, hospitals, professional societies, and foundations and interest groups.

Other organizations also hire scientists for roles in popular science communication. For example, niches sometimes become available at museums and the firms that design exhibits for them, at publishing companies that issue popular science books, and at consulting firms that prepare materials for general and semi-technical audiences.

Freelancing is another alternative. It can, however, be a precarious living. Before relying on it entirely, an individual is wise to have sources of assignments waiting and a strategy for dealing with lean times. One approach is to take a full-time job while establishing contacts, freelancing on the side, and building up a bank account. Another option is to do other work part time and freelance the rest; unlike many freelancers, those with science backgrounds often can find well-paying part-time activity (for example, consulting, technical writing or editing, clinical or laboratory work, or computer programming).

Breaking into Science Communication

Several means are available to increase your skill in science writing, to demonstrate your competence to potential employers, and to test whether work in science communication truly interests you.

One route to a science writing career is formal training—for example, journalism school or a science communication program such as that at the University of California at Santa Cruz. Such education can be particularly valuable as a source of contacts and credibility. A *Directory of Science Communication Courses, Programs, and Faculty* is available through Lawrence Verbit, Department of Chemistry, State University of New York at Binghamton.[88]

Another path is an internship. For example, since 1975, the AAAS Mass Media Science and Engineering Fellows Program has placed up to about twenty advanced science and engineering students per summer at newspapers, magazines, and radio and television stations; many of the participants have gone on to science writing careers. Of course, you may also set up an internship independently.

Freelancing while pursuing traditional scientific activities or while completing your education is still one more route. The published articles ("clips") that you produce can be crucial in convincing editors and others to hire you. An alternative is to take a job in non-science-related journalism (e.g., as a general reporter) to establish and prove your skills.

Another strategy is to work your way from one part of an organization into another or otherwise to create your own niche. For example, you may work mainly in research or administration but sometimes assist a public information office or do some popular science writing as part of your own work. Then presenting science to the public may evolve into a larger and larger part of your role.

Seeking and Accepting Positions in Science Communication

Science-trained science writers are a special commodity. Although they are well suited for a wide range of activities, many employers that could benefit from their services have never considered hiring such individuals or even realized that they exist. Therefore, do not limit your search to advertised positions. If a setting seems like a logical one in which to apply your skills, send a résumé, a cover letter geared to the specific organization, and perhaps some writing samples. Do not give these materials only to the personnel office, which may well dismiss them because you fall outside a standard category. Also submit them to the head of the unit where you may wish to work or to someone else who seems likely to appreciate what you can offer.

In addition, let your pursuit of a science writing job be widely known. Science journalists, science communication faculty, and others often hear of openings and may be able to match you with one. The National Association of Science Writers (P.O. Box 294, Greenlawn, New York 11740) and the American Medical Writers Association (5272 River Road, Suite 370, Bethesda, Maryland 20816) can also be valuable sources of contacts and job information.

Breaking into science communication is usually much harder than changing positions within the field. Therefore, even if a job does not seem ideal, it may be worth taking initially. Once you demonstrate your skill, you will probably find it easy to move sideways and up, especially if you stay alert for opportunities and are vigorous in pursuing them. The biggest logistical hurdle is at the start. Thereafter, the greatest challenges—and the deepest satisfactions—are those of science communication itself.

Recommended Sources

Alberger PL, Carter VL, eds. Communicating university research. Washington, D.C.: Council for Advancement and Support of Education, 1981.
(proceedings of a conference featuring scientists, journalists, and public information professionals)

Bander MS. The scientist and the news media. The New England Journal of Medicine 1983; 308:1170–1173.
(a concise guide to dealing with journalists)

Brady J. The craft of interviewing. New York: Vintage Books, 1977.

Flesch R. The art of readable writing. New York: Collier Books, 1962.

Goodell R. The visible scientists. Boston: Little, Brown and Company, 1977.
(contains a wealth of information not only on the visible scientist phenomenon but also on how the media cover science)

Goodfield J. Reflections on science and the media. Washington, D.C.: American Association for the Advancement of Science, 1981.
(a thoughtful look at science and the press; includes four case studies)

McCall RB, Stocking SH. Between scientists and public: communicating psychological research through the mass media. American Psychologist 1982; 37:985–995.
(contains much material applicable to fields other than psychology)

Miller NE. The scientist's responsibility for public information: a guide to effective communication with the media. Bethesda, Maryland: Society for Neuroscience, 1979. Reprinted by Scientists' Institute for Public Information (New York).
(an outstanding brief guide)

Perlman D. Science and the mass media. Daedalus 1974; 103 (Summer):207–222.

Rogers CL, Dunwoody S. Science, the media, and the public: selected annotated bibliographies. Washington, D.C.: American Association for the Advancement of Science, 1981.

(includes annotated lists of broadcast programs and periodicals featuring science, and of research articles and commentary on science communication)

Sandman PM, Paden M.   At Three Mile Island. Columbia Journalism Review 1979 July/August:43–58.
(a case study of media coverage)

Science, technology, and the press: must the "age of innocence" end? Technology Review 1980 March/April:46–56.
(a discussion with eight prominent science journalists)

Strunk W Jr., White EB.   The elements of style. 3rd ed. New York: The Macmillan Company, 1979.

Verbit LP.   Directory of science communication courses, programs, and faculty. Binghamton, New York: Department of Chemistry, State University of New York at Binghamton, 1983.

Zinsser W.   On writing well: an informal guide to writing nonfiction. New York: Harper & Row, 1976.

References Cited

1. **Anastasi TE Jr.** Communicating for results. Menlo Park, California: Cummings Publishing Company, 1972.
2. . . . And out in space. The Boston Globe 1981 August 27:10.
3. **Asbell B.** The man who built a better mouse. Yankee 1981 December: 190 ff.
4. **Asimov A.** Words of science. Boston: Houghton Mifflin, 1959.
5. **Bander MS.** The scientist and the news media. The New England Journal of Medicine 1983; 308:1170–1173.
6. **Bauer E.** Legibility. In: Reeder RC, ed. Sourcebook of medical communication. St. Louis: C. V. Mosby, 1981:81–86.
7. **Begley S.** What makes a scholarly article newsworthy? Scholarly Communication Around the World: A Joint Global Conference, Philadelphia, 1983 May 16.
8. **Begley S, Hager M.** Nature's tiniest magnets. Newsweek 1982 May 10:46.
9. **Bendiner E.** De Kruif: from practitioner to chronicler of science. Hospital Practice 1979 February:31 ff.
10. **Bennett W, Gurin J.** The dieter's dilemma. New York: Basic Books, 1982.
11. **Beveridge WIB.** The art of scientific investigation. New York: Vintage Books.
12. **Boffey PM.** Forecasters' minor mistakes can produce major surprises. The New York Times 1983 February 14:1, B4.
13. **Brady J.** The craft of interviewing. New York: Vintage Books, 1977.
14. **Brody JE.** Personal health. The New York Times 1983 June 1:C12.
15. **Burkett DW.** Writing science news for the mass media. 2nd ed. Houston: Gulf Publishing Company, 1973.
16. The chemist and the media. Washington, D.C.: American Chemical Society, 1982.
17. **Clark M, Hager M, Gastel B.** Cancer and our diet. Newsweek 1978 July 24:85–86.
18. The court leaves OSHA hanging. Business Week 1980 July 21:67–68.
19. **Crisp D.** Special problems of the scientist and the local press. AAAS Annual Meeting, Washington, D.C., 1982 January 8.
20. **Darrow KK.** How to address the APS. Physics Today 1981 December: 25–29.

21. **Demarest RJ.** Art and illustration. Council of Biology Editors Annual Conference, Boston, 1981 April 28.

22. **Desruisseaux P.** The Ps and Qs of Q and A. CASE Currents 1979 October:32 ff.

23. **Dolnick E.** Physics takes a big step toward unification. The Boston Globe 1983 June 6:42.

24. **Dunwoody S.** The science writing inner club: a communication link between science and the lay public. Science, Technology, & Human Values 1980; 5(Winter):14–22.

25. **Dunwoody SL.** Scientists as sources: what the research shows. AAAS Annual Meeting, Washington, D.C., 1982 January 3.

26. **Dunwoody S.** A question of accuracy. IEEE Transactions on Professional Communication 1982; PC-25:196–199.

27. **Dunwoody S, Patrusky B, Rogers C.** The gatekeepers: the inner circle in science writing. In: Alberger PL, Carter VL, eds. Communicating university research. Washington, D.C.: Council for Advancement and Support of Education, 1981:152–161.

28. **Dunwoody S, Scott BT.** Scientists as mass media sources. Journalism Quarterly 1982; 59:52–59.

29. **Dupont R.** Coping with controversial research. In: Alberger PL, Carter VL, eds. Communicating university research. Washington, D.C.: Council for Advancement and Support of Education, 1981:83–92.

30. The environment: how cancerous? (Meet the Press 1978 June 25.) Annals of the New York Academy of Sciences 1979; 330:799–811.

31. The fairly concise *New Scientist* magazine dictionary. New Scientist 1982 December 23/30:811–813.

32. **Faraday M.** The chemical history of a candle. In: Bowen ME, Mazzeo JA, eds. Writing about science. New York: Oxford University Press, 1979:7–19.

33. **Feder B.** Got a second? Sorry, that's too long. The New York Times 1982 April 18:9E.

34. **Flesch R.** The art of readable writing. New York: Collier Books, 1962.

35. **Frazer J.** Book reviewing at the local level. In: Kamerman SE, ed. Book reviewing. Boston: The Writer, Inc., 1978:99–108.

36. **Freeman E.** Facilitating the transition from scholarly journal to mass media. Scholarly Communication Around the World: A Joint Global Conference, Philadelphia, 1983 May 16.

37. **Funkhouser GR, Maccoby N.** Tailoring science writing to the general audience. Journalism Quarterly 1973; 50:220–226.

38. **Goodell R.** The visible scientists. Boston: Little, Brown and Company, 1977.

39. **Goodell R.** Scientists, not reporters, determined what the press covered in DNA debate. National Association of Science Writers Newsletter 1981 March:1–4.

40. **Goodell R, Russell C.** The scientist as newsmaker. In: Alberger PL, Carter VL, eds. Communicating university research. Washington, D.C.: Council for Advancement and Support of Education, 1981:162–170.

41. **Goodfield J.** Reflections on science and the media. Washington, D.C.: American Association for the Advancement of Science, 1981.

42. **Gould SJ.** The panda's thumb. New York: W. W. Norton, 1980.

43. **Gunning R.** The technique of clear writing. New York: McGraw-Hill Book Company, 1968.
44. **Hammond AL, Lowenberg P, Bazell R.** Serving two masters: journalistic credibility and institutional interests. In: Alberger PL, Carter VL, eds. Communicating university research. Washington, D.C.: Council for Advancement and Support of Education, 1981:179–188.
45. **Hunsaker A.** Enjoyment and information gain in science articles. Journalism Quarterly 1979; 56:617–619.
46. Keep your ear on the ball. Science 83 1983 April:7 ff.
47. **Kidder T.** The soul of a new machine. Boston: Little, Brown and Company, 1981.
48. **Kobre S.** Development of American journalism. Dubuque, Iowa: William C. Brown Publishers, 1969.
49. **Krieghbaum H.** Science and the mass media. New York: New York University Press, 1967.
50. **Leary W, West S, Young P.** The genuine article: reporting real research. In: Alberger PL, Carter VL, eds. Communicating university research. Washington, D.C.: Council for Advancement and Support of Education, 1981:171–178.
51. **Likely A, Kalson D.** Physics goes public. New York: American Institute of Physics, 1981.
52. Living with Alzheimer's. The MacNeil–Lehrer Report 1983 April 12. (Transcript 1967.)
53. **Los Alamos Scientific Laboratory.** Guidebook for news media relations. U.S. Government Printing Office, 1980.
54. **Maclean F.** The query letter. In: Miller L, ed. Questions writers ask. Washington, D.C.: Washington Independent Writers, 1982:3–5.
55. **Marcus SJ.** Comments following lecture by HF Judson. Celebrating and Criticizing Science. (Lecture series.) Massachusetts Institute of Technology 1982 November 16.
56. **McCall RB, Stocking SH.** Between scientists and public: communicating psychological research through the mass media. American Psychologist 1982; 37:985–995.
57. **McDowell E.** Books on science riding wave of popularity. The New York Times 1981 November 28:11.
58. **Merrill PW.** The principles of poor writing. The Scientific Monthly 1947; 64:72–74.
59. **Miller NE.** The scientist's responsibility for public information: a guide to effective communication with the media. Bethesda, Maryland: Society for Neuroscience, 1979. Reprinted by Scientists' Institute for Public Information (New York).
60. **Nelson H.** Some schizophrenia linked to prenatal changes in brain cells. The Boston Globe 1983 June 7:8.
61. **Nunn CZ.** Readership and coverage of science and technology in newspapers. Journalism Quarterly 1979; 56:27–30.
62. **O'Leary D.** Risks and benefits of cooperating with the media. AAAS Annual Meeting, Washington, D.C., 1982 January 8.

63. **Perlman D.** Science and the mass media. Daedalus 1974; 103(Summer):207–222.
64. **Perlman D.** Informing the public about research: the media. In: Alberger PL, Carter VL, eds. Communicating university research. Washington, D.C.: Council for Advancement and Support of Education, 1981:109–115.
65. **Perry S.** 10 "rules" for the media game. The Chronicle of Higher Education 1983 January 19:25.
66. **Plimpton G.** How to make a speech. (Advertisement.) New York: International Paper Company, 1981.
67. **Pollack A.** Lowly beeper coming of age. The New York Times 1982 July 15:D2.
68. **Procter S.** VLBI: fine tuning the universe. Smithsonian Institution Research Reports 1983 Spring:4.
69. **Rathbone RR.** Communicating technical information. Reading, Massachusetts: Addison-Wesley Publishing Company, 1972.
70. **Relman AS.** The Ingelfinger rule. The New England Journal of Medicine 1981; 305:824–826.
71. **Rensberger B.** Science in the media. SIPIscope 1982–83 Winter:15–16.
72. **Rensberger B.** What makes science news? The Sciences 1978 September:10–13.
73. **Rogers CL, Dunwoody S.** Science, the media, and the public: selected annotated bibliographies. Washington, D.C.: American Association for the Advancement of Science, 1981.
74. **Roueché B.** The medical detectives. New York: Truman Talley/Times, 1980.
75. **Russell C, Leary W, Young P, Perlman D.** How I cover science: newspapers. In: Alberger PL, Carter VL, eds. Communicating university research. Washington, D.C.: Council for Advancement and Support of Education, 1981:116–127.
76. **Ryan M, Dunwoody SL.** Academic and professional training patterns of science writers. Journalism Quarterly 1975; 52:239–246, 290.
77. **Sandman PM, Paden M.** At Three Mile Island. Columbia Journalism Review 1979 July/August:43–58.
78. **Saturn/Voyager.** The MacNeil/Lehrer Report 1980 November 14. (Transcript 1340.)
79. **Schemenaur PJ, ed.** 1983 writer's market. Cincinnati: Writer's Digest Books, 1982.
80. Science, technology, and the press: must the "age of innocence" end? Technology Review 1980 March/April:46–56.
81. **Shepherd RG, Goode E.** Scientists in the popular press. New Scientist 1977 November 24:482–484.
82. **Strunk W Jr., White EB.** The elements of style. 3rd ed. New York: The Macmillan Company, 1979.
83. **Sullivan W.** Astronomers find rapidly throbbing pulsar. The New York Times 1983 May 21:24.
84. **Tankard JW Jr., Ryan M.** News source perceptions of accuracy of science coverage. Journalism Quarterly 1974; 51:219–225, 334.

85. **Thomas L.** The lives of a cell. New York: Viking Press, 1974.
86. **Tressel G.** Personal communication.
87. **Tressel G, Goodell R, Moss TH, Stockton W.** How the public views science and research. In: Alberger PL, Carter VL, eds. Communicating university research. Washington, D.C.: Council for Advancement and Support of Education, 1981:7–26.
88. **Verbit LP.** Directory of science communication courses, programs, and faculty. Binghamton, New York: Department of Chemistry, State University of New York at Binghamton, 1983 (in press).
89. **Wade N.** Medical journal draws lancet on rival. Science 1981; 211:561.
90. **Walsh J.** Public attitude toward science is yes, but—. Science 1982; 215:270–272.
91. **Weiner J.** Prime time science. The Sciences 1980 September:6–11.
92. **Zinsser W.** On writing well: an informal guide to writing nonfiction. New York: Harper & Row, 1976.

Index